the Hawkweed
PASSIVE SOLAR HOUSE BOOK

the Hawkweed
PASSIVE SOLAR HOUSE BOOK

by The Hawkweed Group

Rodney Wright
Sydney Wright
Bob Selby
Larry Dieckmann

Rand McNally & Company
Chicago • New York • San Francisco

PHOTO CREDITS

George Fred Keck and William Keck, F.A.I.A., 8, 9; Hedrich-Blessing, 10, 53, bottom; The Newberry Library, Chicago, Illinois, 50, top; Courtesy of *Daily Courier News,* Crystal Lake, Illinois, 85; Courtesy of The Art Institute of Chicago, 121.

All other photographs were taken by The Hawkweed Group.

Library of Congress Cataloging in Publication Data:

THE HAWKWEED PASSIVE SOLAR HOUSE BOOK.
Bibliography: p.
Includes Index.
1. Solar houses—Design and construction.
2. Solar energy—Passive systems. I. Wright,
Sydney.
TH7414.H38   690′.869   80-20932
ISBN 0-528-81107-X
ISBN 0-528-88034-9 (pbk.)

First printing, 1980

**Page 2:
Sunspace and
greenhouse under
construction**

Contents

Foreword

"Again when [Socrates] said about houses that beauty and utility were the same, he was giving a lesson in the art of building houses as they ought to be. He approached the problem thus: 'Is it not true that to have the right sort of house it should be both useful and pleasant to live in?'

"And this being admitted, 'Is it pleasant,' he asked, 'to have it cool in summer and warm in winter?'

"And when they agreed with this also, 'Now in houses with a southern exposure, the sun's rays penetrate the porticoes in winter, but in summer, being less inclined, they afford us shade. If, then, this is the best arrangement, we should build the south side loftier to get the winter sun, and the north side lower to keep out the cold winds.' " (Xenophon's *Memorabilia*, Book III, Chapter 8. As translated from the Greek by Nino Repetto.)

When the Chicago-based office of Keck and Keck began to pursue the use of solar energy to help heat buildings in the 1930s, we found that there is nothing new under the sun.

In our hometown of Watertown, Wisconsin, for example, there is a 125-year-old octagonal house, which was designed not specifically as a solar house but as a way of making life easier in the settlement of that portion of the state. In its design it solved a number of problems in the use of energy; if these techniques had been developed during the later growth of domestic architecture in this country, they would have given much more impetus to the creation of adequate energy-efficient housing.

The Octagon House, which today is owned by the Watertown Historical Society and is open to the public for tours during the

7

Octagon House, Watertown, Wisconsin (1852), with original porches still in place; they have since been removed.

tourist season, was designed and built in 1850–52 by a lawyer who wished to settle in the Midwest, with a promise to his future wife of "the best house in that part of the country." He came from the East and undoubtedly had seen this type of house in that area. His analytical mind told him that the greatest amount of enclosed space with the least amount of external area for heat loss would be a circle. He approached this by choosing an octagonal shape, which was simpler to construct. The brick he used came from Milwaukee, a distance of 45 miles, and had to be hauled over plank roads by ox cart, since the railroads did not reach Watertown until 1855 or '56.

To provide running water, the house was built with a flat roof to collect rainwater, which flowed to an ingenious reservoir on the top floor of the building, made up of wood with wedges that could be driven tight in case of an extended drought. An overflow brought the water to a masonry cistern in the basement adjacent to the kitchen. The excess water from the cistern was wasted down the side of the hill. A lead-pipe plumbing system provided running water with a line from the reservoir down to a coil in the kitchen stove and/or the heating plant, running back up to outlets in the stair hall as well as to a zinc-lined bathtub on the second floor.

A central heating plant was installed in the basement under the circular stairway. This stair ran only from the first floor up to the cupola at the roof level. A brick enclosure formed the housing in the basement, within which was located a cast-iron burner capable of taking 4-foot-long cordwood. Ducts were run in the hollow brick walls of the staircase up to all the principal rooms of the

House of Tomorrow, built for the 1933 Chicago World's Fair by Keck and Keck.

house. At the rear of the housing was an access door through which pans of water were placed to help humidify the heated air.

The centrally located circular stairway provided an excellent ductwork system to ventilate the house in the summertime. One had only to open the windows of the cupola on the lee side and open the windows on the windward side of the house to create a natural vent duct acting by gravity to bring in cooler air, especially at night, without any mechanical ventilation.

The porches and roofs on the first two floors (now removed) provided the necessary sun control in the summer to help keep the house cool, and the shutters on the windows contributed to maintaining a warmer house in the winter, acting as storm sash.

In short, the house was well thought out to provide greater comfort to its inhabitants by taking advantage of natural phenomena.

In 1933, eighty years after the construction of the Octagon House, our office unintentionally built its first solar-heated house by designing a building for A Century of Progress, the Chicago World's Fair. Named the House of Tomorrow, this house was designed to show the maximum of progress available in construction technique and mechanical equipment. It was 12-sided (duodecagonal), with a central circular stairway around a utility core. Workshop, garage, recreation room, utility room, and an airplane hangar were on the ground floor; living and bedroom areas were on the first floor, with a conservatory above.

While the house was under construction in February, 1933 (before the heating plant was in operation, but with the house enclosed against the elements), my brother George Fred Keck visited the building on a sunny day with the temperature hovering between zero and 10° Fahrenheit. Upon entering the living room, he found carpenters and other workmen laboring in their shirt sleeves. Surprising as this seemed at first, when you stop to think about it, it was known that any greenhouse operator would bank his boiler in the morning of a winter day when the sun was shining, turning it on again only when the sun was gone for the day.

Based on the above observation, our office tried to find more information on solar energy. None was available, in addition to which, the mechanical engineers at that time stated that it would be too costly to heat such a house and that it would not be comfortable to live in. Not until 1940 was our office able to find actual data on solar heat gain.

Since that time, our office has built more than 300 homes that are at least partially heated by passive solar application. You might think of them as belonging to the "first generation" of solar homes built in this century.

It was during those early years of the firm that the term "solar house" was invented. In 1940 we built a home for developer Howard Sloan of Glenview, Illinois, which incorporated our current solar thinking. Al Chase, the real estate editor of the *Chicago Tribune*, visited the building and called it a "solar house," which we believe is the first time this term was used.

Sloan House, Glenview, Illinois (Keck and Keck, 1940), possibly the first in America to be called a "solar house."

These homes were far from reaching the energy-efficient levels of today's solar houses. One must remember that until the energy crisis of the 1970s (when the price of fuel escalated), it was uneconomical to provide more than 3 or 4 inches of insulation in the walls and ceilings. Beyond that point the savings of fuel on the additional investment would not pay for the extra cost.

The increase in costs of energy has made it necessary to find ways and means of conserving energy as well as to find alternate sources. This becomes the job of the younger generation.

One of the leading solar architects working today, in my opinion, is Rodney Wright, president of The Hawkweed Group of Chicago, Illinois, and Osseo, Wisconsin. I have known Rod for about two decades, and my office has worked closely with him on certain projects.

Rodney's outstanding contribution to the profession has been his research and application of solar principles in the residential and commercial fields of construction. One cannot overrate this point in view of the present energy crunch! Some of his recent work has been approaching the 100% mark in energy efficiency in the rugged climate of the Midwest. This is a remarkable accomplishment.

I was pleased to learn that The Hawkweed Group was planning this book. The text of the book explains passive solar techniques to the general public in simple, understandable terms, and carries on the tradition of the passive work that my brother and I initiated, a subject that will be in everyone's future as time goes on. It is mandatory that further work be done, with information disseminated to the public, to find ways of replacing our diminishing supply of fossil fuels with this great renewable source—the sun. We must learn to harness the energy of the sun in the best possible manner.

William Keck, F.A.I.A.
January, 1980

The Newcomer House of
Bryan, Ohio.

Preface
and Acknowledgments

The authors of this book are The Hawkweed Group. Three architects and a city planner, we have been in practice since 1960. Our very first solar house was designed and built in 1961, in northern Illinois, and the original owners still live in it. In 1973, we decided to do only solar projects. At that time we changed the name of our firm, borrowing that of a flower—*Hieracium Aurantiacum,* the orange hawkweed or devil's paintbrush, a member of the Composite family and a common plant in upper Wisconsin. This flower, like the sunflower, tracks the sun and closes itself up at night to hoard its store of energy.

Since 1973, we have designed over 300 units of solar buildings: residences, a motel, fire stations, a recreation center, a grocery store, a medical center, an office building, maintenance buildings, housing for the elderly, municipal buildings, a bathhouse, town houses, a warehouse, vacation homes, a gas station. We have done new buildings and we have done retrofit (re-fitting for solar heat) for existing buildings. We have designed one-story buildings and buildings of two stories or more, with basements and without. Some are earth-integrated buildings, that is, they are dug into hillsides or have earth pushed up around them after construction. Some have wood-burning stoves as a second source of heat, thus using the sun in two ways—directly, in solar heating, and indirectly in the stored solar energy of the wood that they burn.

The construction materials we use for solar houses are commonplace: wood, concrete block, concrete, and brick. The

Built by a member of The Hawkweed Group for his own family, this free-standing house has a solar roof for heat collection.

locations of our houses are all over the Midwest and in a few other places besides: Illinois, Minnesota, Ohio, Wisconsin, Indiana, Michigan, North Carolina, and Colorado. You can build solar buildings wherever you wish—and we have done so.

In addition, several of us are erecting our own solar houses on our farm in Wisconsin. One of us is retrofitting a 100-year-old farmhouse, a second is building a new free-standing house—small, two-story, and without interior walls to permit free movement of heated air. The third house sits on a recycled barn foundation, is earth-integrated, and, like the others, has an airtight wood stove for backup heat and cooking.

We also work with the sun to grow our own food in our Wisconsin gardens. Vegetables from the garden are better than money in the bank, we think. We can, store, and freeze enough to last the winter. Greenhousing will soon bring us a solar solution to fresh vegetables year round. As a result of our gardening experience, we often help our clients to plan their garden and orchard space.

Hoping to pass on what we have learned about solar usage—from climate analysis to house design, with a great deal in between—we teach seminars and workshops, give speeches and interviews, and write papers and articles. Out of these activities grew this book, which is our attempt to share with many people at the same time our blend of the theoretical and the practical, and to help in the birth of the solar age.

This book, like all of our work, is a team project, and all of us were involved in its writing. Rodney Wright, architect, founder and president of The Hawkweed Group, is our silver-haired leader and the designer of our first solar house. Bob Selby, architect, took most of the photographs and did many of the drawings for the book; Sydney Wright, city planner, sorted out all of our ideas and

Solar retrofit of a 100-year-old farmhouse.

contributions and put the words together; Larry Dieckmann, architect, used his skill with our computer to provide many of the technical details. Weston Wright, apprentice and university student, drew most of the floor plans.

We would like to thank some of our solar friends: Grace and Gene Theios, who live in our first solar house, built when it was simply the right thing to do; Estelle and Bruce Von Zellen, for believing in the solar solution many years ago; the Keck brothers—William and George Fred—for their encouragement and inspiration; Joe Yohanon, for letting us teach the solar story at Harper College, when people first began listening; Libby Hill, who attended our seminar at Harper and has, through the environmental-action group Epoch B, broadly shared solar facts with people throughout the Chicago area; Bethe Hagens and Jim Laukes, for sharing their publication, *Outlook,* and their amazing ideas with us; Kathy Fairchild and Diane Fahnstrom, for endless typing and detail-checking; John Schaefgen, landscaper, an extraordinary architect of the natural environment; all of our clients, for their parts in making this book happen.

An extremely special thanks to the SUN, without which this book would not have a ray of hope.

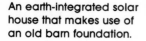

An earth-integrated solar house that makes use of an old barn foundation.

the Hawkweed
PASSIVE SOLAR HOUSE BOOK

Introduction

Imagine your home heated by the sun. Using little energy, heated air moves through the house, making it comfortably warm even on the coldest winter days. In summer, this home will remain comfortably cool when temperatures are high outdoors. During daylight hours, daylight rather than artificial lighting will illuminate the rooms—even on cloudy days.

A passive solar house functions in exactly this way, using renewable energy and renewable materials in construction of the house, while reducing dependence upon fossil fuels. As architects, we have been designing such buildings since 1961, and since 1973 we have designed only solar-heated buildings.

In this book we wish to dispel the three myths of solar heating and cooling—that solar heating is unsuited to cold climates, that it is not yet a mature technology or is too complex, and that it is much too costly for the average homeowner. In fact, nothing could be further from the truth, as we will show you.

We will begin with a discussion of climatic conditions. Temperature, Degree Days, wind patterns, and incidence of sunny, partly cloudy, and cloudy days all form part of the important climatic background for the solar-heated house. As you will see, a solar house is a climate-responsive house. To illustrate this principle, we will describe historic responses to climate in design of shelters, from the time of early American Indian cultures to the 20th century. Then we will detail how to build a climate-responsive house today, incorporating a high degree of thermal efficiency, and show how, even in the severe climate of the upper Midwest, the sun can satisfy 80% to 90% of your annual space-heating requirements.

There are various methods for solar heating a building, and these may be divided into two categories. Some systems do indeed use high technology, requiring complex instrumentation and expensive equipment, such as manufactured flat plate or concentrating solar collectors applied to the finished roof. But others are low-technology methods, using readily available materials and construction techniques that are normal to building trades. Solar technology is integrated into the architectural design so that *the building itself* is the heating system, constructed of materials that effectively capture, store, and distribute the sun's heat. The solar-heating system may be as simple as south-facing glass or a glazed masonry south wall.

Any of the low-technology, architecturally integrated systems can be defined as passive, and these passive systems are the subject of this book. Using the single-family residence as our example, we will show you the entire process of design and construction of a new passive solar building, or a solar addition to an existing one. We will try to enlarge your understanding of passive techniques, helping you decide what is most suitable for your personal needs, your site, and your climate.

We will show you how to plan your house so that it will use as little energy as possible, not only for space heating, but also for cooling and for lighting, and we will give you methods of calculating the necessary size of the system. You will find that the information ranges from the disarmingly simple (face the building south) to the distressingly complex (calculate your heat loss). This is a marvelous illustration of an essential truth: Use of solar heat is a simple matter that you can make very precise by the correct application of techniques and calculations.

We use the Midwest as the base for much of our data and for most of our examples. There are three reasons for this. First, the most severely demanding winter conditions in the lower 48 states are in the Midwest—if you can build a solar house here, you can build it anywhere. The Midwest is the test, and solar usage passes with high marks. Second, this is the center of our professional lives, and the greatest number of examples of our work are sited in the Midwest. The last reason is that the 12 states of the Midwest have more than one-fourth of this country's total population. If this region turns to the sun, it will be possible to severely prune the fossil-energy budget for the entire nation.

Don't worry, though, if you don't live in the Midwest, since we will show you how to adapt the principles described to your climate conditions, wherever you live.

Then we will outline the design and building processes, supplying information that will assist you in working with an architect or contractor—or that will enable you to be your own designer or do your own construction, if you are already skilled in construction techniques. Here we will address the myth of cost. As we will show you, your solar house, designed and built as an integral system, will be no more expensive to build and less expensive to maintain than a nonsolar house of the same size and type. The simplicity of its design, the integration of the solar-heating components as a part of the building itself, replacing other materials in roof, walls, and floor, brings the cost within the normal range.

As we provide you with all of this information about design and construction of your solar house, we will show you some of the buildings that we have designed, using examples that we think will help you to understand our concepts. We will also tell you how one lives in a solar house. This is a new experience for all of us, since there are shutters to open and close, dampers to check, and details of all sorts to manage and watch. This is a house for interaction, not for button pushers.

In the last section, we will describe some large-scale community projects that are pointing the way to our solar future and share with you our thoughts on how land-use planning for solar usage can reduce our total energy consumption and contribute to the greenhousing of America.

Ever since we designed our first passive solar house in 1961, we have become increasingly convinced of the integrity of these buildings. The sun, after all, is the only heat source that is freely available. It is not subject to embargo or disruption of supply. The price of sunshine cannot be manipulated by giant corporations and will not fuel inflation. Sunshine does not require waste-disposal sites, permanent guards, or lead shields. No one must fear contamination from integrated solar buildings, nor will these buildings need to be mothballed to protect future generations from radioactivity or other side effects. The sun will just go on warming our bodies, brightening our lives, and saving other forms of energy.

The solar resource, we believe, best serves the needs of an egalitarian, democratic society. It is time to get on with the business of building a solar democracy.

Analyzing
Climatic Conditions

1

The gardens that we plant on our Wisconsin land are a continuation of midwestern tradition. The Garden is what the Midwest was called when it was settled, and it is still a garden. With deep topsoil, plentiful moisture, and abundant sun, the land is highly fertile.

The Midwest conjures images of the vast prairie that once extended from Minnesota to Missouri and from Indiana to Kansas. Heavy forest intruded into the prairie on the east and north, and the dry upland, or steppe, formed the western boundary. Long summer days, spring rains, winter snows—all influenced by the sun—insured good crops. Without this solar beneficence, rapid settlement and agricultural development would have been difficult. Today, the 12 fertile states of the Midwest—Illinois, Indiana, Iowa, Kansas, Michigan, Minnesota, Missouri, Nebraska, North Dakota, Ohio, South Dakota, and Wisconsin—grow a whopping 40% of the country's food.

Just as climate influenced the settlement and enhanced the prosperity of the Midwest, so climate influences decisions about building design. Let's examine three major climatic elements—solar radiation, temperature, and wind patterns—first seeing how they function and then looking at their significance for a solar house. These climate elements are all part of the ''macroclimate,'' or the overall weather conditions in a given area. The peculiarities and variabilities of local climate conditions at the building site—the ''mesoclimate''—will modify or exaggerate the macroclimate. Finally, there are climatic conditions at work within

22

Sunlight filters through bare trees in winter.

the house itself, known as the "microclimate." Only by studying all three types of climate—macro, meso, and micro—can you get a full picture of the effects of climate on you and your house.

Macroclimate

Solar Radiation. Without that constant, renewable energy source called the sun, all other climate events would not occur, nor would life as it is known on this planet. Although less than one five-billionth of the sun's energy is intercepted by the earth, this energy generates our weather, creates the circulatory patterns of our atmosphere and oceans, and is the ultimate source for all of the energy we use on earth.

Not all of the solar radiation beamed at the earth reaches its surface. As it travels through the atmosphere, for example, some of the energy will be absorbed by the atmosphere, some will be reflected or scattered by dust or clouds. Less strong, it is this diffuse sunlight that provides light on cloudy days. And a certain amount of the radiation that does reach the surface will not be absorbed but will be reflected by such things as water or snow.

In addition to atmospheric conditions, two other factors affect the amount of solar energy that reaches the earth, intensity and duration. The intensity of solar radiation is determined by the angle of the sun's rays as they touch the earth. In summer, when the tilt of the earth's axis points the Northern Hemisphere toward the sun, this angle is more nearly vertical to the earth's surface, increasing the intensity of solar energy received. In winter, the earth's tilt is away from the sun in the Northern Hemisphere and the rays of the sun reach the surface at a more oblique angle. This means that the radiation is spread over a larger segment of the earth and must pass through more of the atmosphere as it travels to the surface. Therefore, more energy is lost by scattering, reflectance, and absorption.

The path the earth follows around the sun affects the amount of solar radiation received on the earth's surface.

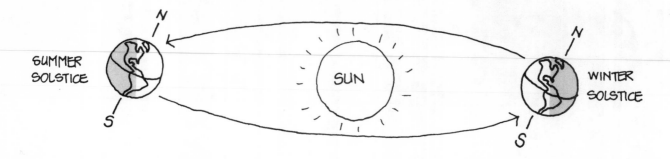

SUMMER SOLSTICE   SUN   WINTER SOLSTICE

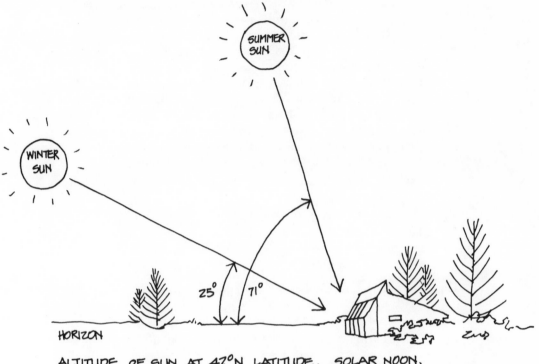

ALTITUDE OF SUN AT 42°N LATITUDE, SOLAR NOON,
WINTER SOLSTICE AND SUMMER SOLSTICE

The duration of solar energy is the number of hours of daylight during which solar radiation is available. On about March 21 and September 23, at the spring and fall equinoxes, the sun at the equator is vertical to the earth. At these times, each 24-hour period is equally divided between day and night. On about June 22, in the Northern Hemisphere, the summer solstice occurs. At this time, midway between the equinoxes, the Northern Hemisphere is tilted toward the sun to its greatest limit, causing the longest day of the year. The opposite happens on about December 22, when the Northern Hemisphere reaches its maximum tilt away from the sun and the longest night of the year occurs. At 40° north latitude on the December solstice, there are nine hours, 20 minutes of sun, and on the summer solstice there are 15 hours of sun, a difference of five hours, 40 minutes.

We have just shown that there is more solar radiation in summer than in winter, because of the angle of the sun and the length of the day. In actual fact, the amount of sunshine received during a given time of year can be well under the potential maximum. Take Chicago, for example. According to a study published by the American Institute of Architects, during any given year this city has an average of 117 clear and sunny days, 120 partly cloudy days, and 128 cloudy days. Translated into hours of sunshine per year, this means that Chicago has 2,611 out of a possible 4,490 hours of sunshine, or sunshine 59% of the potential time. During the three coldest months of the year— December, January, and February, which represent 55% of the total annual heating load—the average sunshine figures are:

December—7 clear days, 8 partly cloudy days, 16 cloudy days—115 sun hours

January—8 clear days, 9 partly cloudy days, 14 cloudy days—131 sun hours

February—7 clear days, 8 partly cloudy days, 13 cloudy days—147 sun hours

This means that in December there is sunshine 40% of the time during daylight hours; in January, there is sunshine 45% of the time; and in February, 49% of the time. Although these percentages may seem low, we know from our experience that the

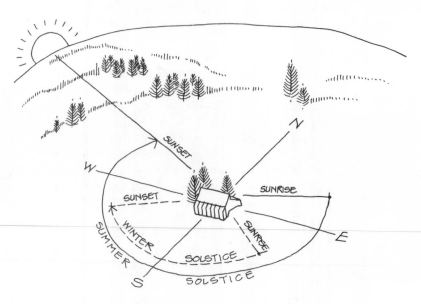

ARC TRACED BY THE SUN BETWEEN SUNRISE AND SUNSET, 42°N LATITUDE, WINTER SOLSTICE AND SUMMER SOLSTICE

Annual Percentage of Possible Sunshine

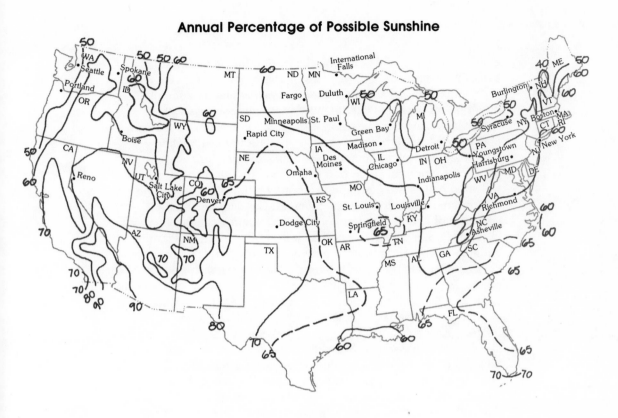

Annual Heating Degree Days
(× 1,000)

amount of sun available in Chicago during the winter months is sufficient to provide 90% or more of space-heating requirements.

Temperature. Seasonal temperature changes are brought about by the seasonal variation in amount of solar energy received. In summer the amount of radiation available increases and so does the temperature, while the amount of radiation available and the temperature are reduced in winter. There is a delay of a month or more before the high or low swing of solar radiation makes itself felt in corresponding temperature change. In the continental United States, generally, January will be the coldest month and July the hottest. Near the oceans or other large bodies of water, the delay in change will be greater, so that the extreme temperature months may be February and August.

In order to find out what temperature means to the design of a house, temperature information must be turned into a measure of heating need. This measure, called a Heating Degree Day, is defined by the National Oceanic and Atmospheric Administration (NOAA) as the number of degrees the daily average temperature is below 65° Fahrenheit. Normally heating is not required in a building when the outdoor average daily temperature is 65°. Heating Degree Days are determined by subtracting the average daily temperatures below 65° from the base number 65. Thus a day with an average temperature of 50° has 15 Heating Degree Days, while one with an average temperature of 65° or higher has none. In this and following chapters, we will be referring to Heating Degree Days as simply Degree Days (DD). In the continental United States, the number of Degree Days per year ranges from about 10,000 near the Canadian border to around 100 in the Florida Keys.

Winds. Winds have two important effects upon climate: Through the process known as advection, they move heated air toward cooler air, and they carry water vapor from the oceans to land masses, where it condenses and falls as rain or snow. Winds in the Northern Hemisphere are strongly affected by large land masses and, as a result, are complicated systems, subject to behavior that is difficult to predict. However, it is generally true that winter winds are strong and typically come out of the west, while summer winds are mild and come from various directions, but most frequently out of the south in the upper Midwest.

The importance of wind data lies in the ability of the wind to carry heat from warmer areas to cooler ones. Thus winter wind steals heat from the house, while summer wind helps to cool it.

At the end of this chapter, we present a weather summary for

31 representative cold-climate cities around the country. By selecting the one closest to you, you can get an idea of the general conditions in your area. You can obtain more specific climate information for many localities in the United States by writing to NOAA at the National Climatic Center, Federal Building, Asheville, North Carolina 28801. The *Climatic Atlas of the United States* provides a great deal of information. There are also bulletins for many cities, known as *Local Climatological Data,* which provide summaries of climate conditions for each of 291 reporting weather stations.

Mesoclimate

When it comes to finding out about the climate conditions in your town and at your building site, you will have to become an amateur meteorologist yourself. First, if you do not live near an NOAA weather reporting station, you will need to interpolate from nearby stations, perhaps using more than one set of data. You can check your assumptions against any local information that you can find, from newspaper records, amateur weather watchers, local airports and pilots, local schools and colleges, and longtime

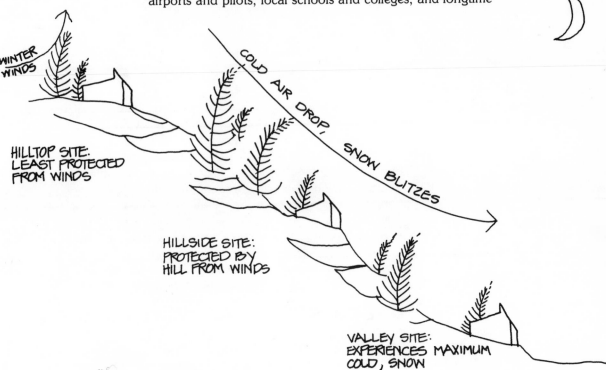

WINTER WINDS

COLD AIR DROP, SNOW BLITZES

HILLTOP SITE: LEAST PROTECTED FROM WINDS

HILLSIDE SITE: PROTECTED BY HILL FROM WINDS

VALLEY SITE: EXPERIENCES MAXIMUM COLD, SNOW

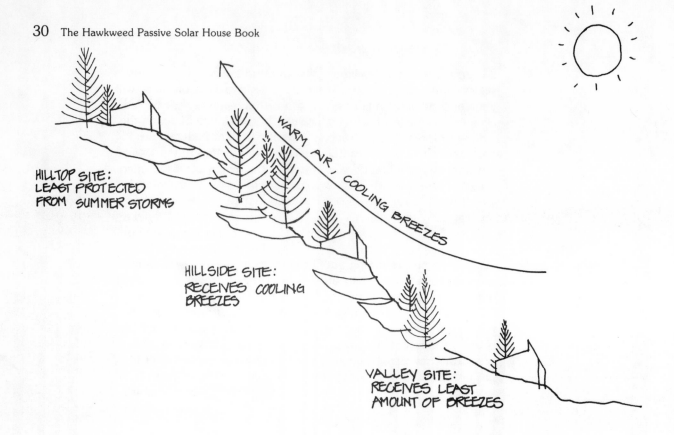

HILLTOP SITE:
LEAST PROTECTED
FROM SUMMER STORMS

WARM AIR, COOLING BREEZES

HILLSIDE SITE:
RECEIVES COOLING
BREEZES

VALLEY SITE:
RECEIVES LEAST
AMOUNT OF BREEZES

residents. Some of these sources may have useful weather records.

The next step is to analyze the immediate area for features that will modify the climate. If your building site is in a valley, cold air will drain down the valley slope in winter, making the bottom of the valley the coldest place around. Valleys may also generate winds, intensifying winter cold. A snow blitz may pour down the valley in winter, dumping more snow there than on the slopes.

On the slope of a hill, breezes will generally move up in summer, providing a cooling effect. Sun will be available on southern slopes, but on northern slopes there will be no solar heat gain. Eastern and western slopes will receive some sun. At the top of the hill, there is greatest exposure to winds. This means an increase in winter cold and an increased heating requirement for a building.

Woodlands can influence the mesoclimate in several ways. The temperature within a wooded area is lower by ten degrees or more in summer than adjacent open space. This affects temperatures nearby, lowering them slightly as cooler air is drawn from the woods. Woodlands also block winds, and they could block the sun from an adjacent building site. All types of plant cover, in fact, will alter climate conditions, because air moving over plants is cooled.

A body of water may also have a moderating influence on the surrounding area, depending upon its size and location. Water temperatures tend to remain more constant year-round than do air and land temperatures. Air flowing over the water will be warmed in winter and cooled in summer, thus somewhat modifying local temperature levels.

Although you need not be a total expert on climate to build a solar house, the more you understand it, the better equipped you are to adjust to it. Two excellent books that provide detailed information on the analysis of mesoclimate conditions are *Design with Climate* by Victor Olgyay and *The Climate Near the Ground* by Rudolf Geiger.

Microclimate

In any discussion of climate, the most important type for you is personal living climate—the microclimate of your house. This is the area where you will feel too hot or cold, too dry or damp, shivering, at ease, just right, pleasant, comfortable. Here you will have a good or bad disposition because of warm or cold walls or floors, cool or warm air flow.

You want a space where it is comfortable for you to cook, sleep, rest, work, and play. Human comfort is affected by temperature, wind, relative humidity, and solar radiation. The human body generates energy and must continuously lose some of its heat to maintain a balance. Cold and windy conditions can cause heat loss that is too great to be made up, resulting in uncomfortable chilliness. In hot weather the body is unable to lose enough heat, which is also uncomfortable. High relative humidity makes hot weather feel even hotter, since heat cannot be lost by evaporation of body moisture.

It is generally recognized that the most comfortable temperature for the human body is between 65° and 82° Fahrenheit, the exact point depending upon such conditions as humidity and air circulation, and that no heating or cooling is required within this range. The amount of relative humidity that is tolerable varies with the temperature. At a temperature of 82° a relative humidity above 45% is not comfortable, but at 70° relative humidity can be as high as 75% without discomfort.

Given an understanding of the body's needs, climatic conditions can be modified within the house to provide for human comfort. On a cold day in winter, with the sun shining, you will be warm—perhaps even hot—when in a greenhouse or at a south-facing window. This natural energy, warming your body and your house, will provide winter comfort on a cold day.

If interior surfaces of the house, such as walls and floors, are cool, your body will lose heat to these surfaces and you will feel cold, even in a room with an air temperature as high as 80°. In the same way, if you are near a window in winter, you will feel colder, because the heat in the room will be moving directly through the

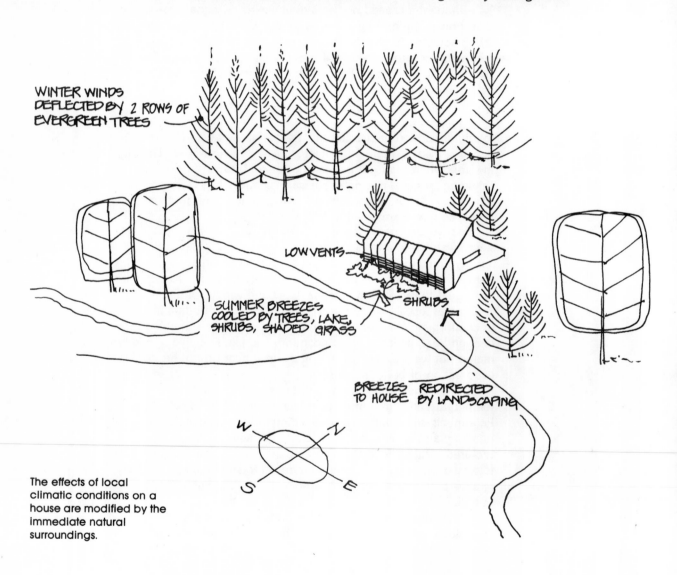

WINTER WINDS DEFLECTED BY 2 ROWS OF EVERGREEN TREES

LOW VENTS

SHRUBS

SUMMER BREEZES COOLED BY TREES, LAKE, SHRUBS, SHADED GRASS

BREEZES REDIRECTED TO HOUSE BY LANDSCAPING

The effects of local climatic conditions on a house are modified by the immediate natural surroundings.

glass to the cold outside, unless the sun is shining through that window to warm you. Poorly fitted doors and windows, holes and cracks in the outside surfaces of the house will permit warm air to escape and cold air to come in.

By contrast, in a well-insulated solar house, the interior surfaces of the walls will be warm—maybe warmer than the air in the house. Double or triple glazing (that is, double or triple panes of glass) and insulating shutters will reduce loss of heat through windows. Weather stripping, air locks (an earlier generation called them vestibules), and a good vapor barrier will help to eliminate drafts. In such a house, the air temperature can be much lower—around 65°—and you will still feel comfortable.

In the summer, you will want to feel cool even on hot, sticky, humid days, when you are working, resting, or sleeping. Air movement has a cooling effect on the human body, removing both moisture and heat. Shading of the house from summer sun will help to keep interior temperatures down. And so, in the shade of friendly trees, with generous overhangs on the south side of the house and windows open to direct breezes, you will have the first ingredients for naturally cooling your house in summer.

Selected Climate Summaries

In order to design a passive solar house, then, you must find out what the climate is like where you plan to build it. Expressed in the terms used in this chapter, you must find out the effect of the macroclimate on the microclimate, through the intervention of the mesoclimate. You will need to know how long the heating season is, how many Degree Days there are, the amount of solar radiation available, the annual temperature range, the relative humidity, and summer wind speeds. With this information, expressed in averages over a period of years and combined with mesoclimate information, you will be prepared to determine how much of your heating can be done by the sun and how much of your summer cooling can be done naturally.

Following are analyses of the macroclimates of 31 cities with varying climatic conditions, in different parts of the country where heating is a prime requirement. The data are taken from the two NOAA publications cited above. Each analysis specifies what part of the total heating need can be met by solar heating, and to what extent the cooling need can be met by shading and ventilation. These summaries do not include particular mesoclimate information—that, as we have said, is for you to find out on your own.

This solar house will
receive heat through a
mass heat wall on its
south face, while the rest
of the house is buttoned
up against the climate.

Analyses of 31 Selected Macroclimates in the United States

Asheville, North Carolina

Latitude: 36°34' N
Sun hours/year: 2,646
Degree Days/year: 4,237
Avg. wind speed: 6 mph

Heating need: Heating is required from September through June, with 55% of the total heating need occurring in December, January, and February. Normal low in January is 27°F, with extreme lows to −7°F.

In a thermally efficient house, solar heating can provide 99% of the total heating requirement, with the solar contribution ranging from 99% in January to 100% February through November.

Cooling need: Relative humidity and temperature are above the comfort range part of the time in July and August, with humidity averaging 96% and temperature averaging a high of 84°F. Extreme high temperatures to 96°F.

In a thermally efficient house, cooling needs can be satisfied naturally by a combination of shading from summer sun and winds admitted to the building at a speed of 3.5 miles per hour.

Boise, Idaho

Latitude: 43°37' N
Sun hours/year: 3,006
Degree Days/year: 5,833
Avg. wind speed: 8 mph

Heating need: Heating is required from August through June, with 50% of the total heating need occurring in December, January, and February. Normal low in January is 21.4°F, with extreme lows to −23°F.

In a thermally efficient house, solar heating can provide 95% of the total heating requirement, with the solar contribution ranging from 89% in January to 100% February through October.

Cooling need: Relative humidity and temperature are above the comfort range part of the time in July and August, with humidity averaging 34% and temperature averaging a high of 90.5°F. Extreme high temperatures to 111°F.

In a thermally efficient house, cooling needs can be satisfied naturally by a combination of shading from summer sun and winds admitted to the building at a speed of 8 miles per hour. However, winds at this speed may cause discomfort to residents simply because of the rapidity of air movement.

Boston, Massachusetts

Latitude: 42°22′ N
Sun hours/year: 2,615
Degree Days/year: 5,621
Avg. wind speed: 11 mph

Heating need: Heating is required from August through June, with 54% of the total heating need occurring in December, January, and February. Normal low in January is 22.5°F, with extreme lows to −12°F.

In a thermally efficient house, solar heating can provide 93% of the total heating requirement, with the solar contribution ranging from 88% in January to 100% April through November.

Cooling need: Relative humidity and temperature are above the comfort range briefly in August, with humidity averaging 58% and temperature averaging a high of 81.4°F. Extreme high temperatures to 102°F.

In a thermally efficient house, cooling needs can be satisfied naturally by a combination of shading from summer sun and winds admitted to the building at a speed of 1.5 miles per hour.

Burlington, Vermont

Latitude: 44°28′ N
Sun hours/year: 2,178
Degree Days/year: 7,876
Avg. wind speed: 8.8 mph

Heating need: Heating is required 12 months of the year, with 52% of the total heating need occurring in December, January, and February. Normal low in January is 7.6°F, with extreme lows to −30°F.

In a thermally efficient house, solar heating can provide 92% of the total heating requirement, with the solar contribution ranging from 83% in December to 100% May through October.

Cooling need: At no time are temperature and relative humidity above the comfort range.

In a thermally efficient house, cooling needs can be satisfied naturally with shading from summer sun.

Chicago, Illinois

Latitude: 41°47′ N
Sun hours/year: 2,611
Degree Days/year: 6,127
Avg. wind speed: 10.4 mph

Heating need: Heating is required from September through June, with 55% of the total heating need occurring in December, January, and February. Normal low in January is 17°F, with extreme lows to −19°F.

In a thermally efficient house, solar heating can provide 93% of the total heating requirement, with the solar contribution ranging from 89% in December to 100% April through November.

Cooling need: Relative humidity and temperature are above the comfort range about 25% of the time in July and August, with humidity averaging 55% and temperature averaging a high of 84°F. Extreme high temperatures to 104°F.

In a thermally efficient house, cooling needs can be satisfied naturally by a combination of shading from summer sun and winds admitted to the building at a speed of 3.5 miles per hour.

Denver, Colorado
Latitude: 39°45′ N
Sun hours/year: 3,033
Degree Days/year: 6,016
Avg. wind speed: 8.5 mph

Heating need: Heating is required from September through June, with 50% of the total heating need occurring in December, January, and February. Normal low in January is 16.2°F, with extreme lows to −30°F.

In a thermally efficient house, solar heating can provide 99% of the total heating requirement, with the solar contribution ranging from 98% in January to 100% April through November.

Cooling need: Relative humidity and temperature are above the comfort range briefly in August, with humidity averaging 36% and temperature averaging a high of 87.4°F. Extreme high temperatures to 104°F.

In a thermally efficient house, cooling needs can be satisfied naturally by a combination of shading from summer sun and winds admitted to the building at a speed of 2.3 miles per hour.

Des Moines, Iowa
Latitude: 41°32′ N
Sun hours/year: 2,770
Degree Days/year: 6,710
Avg. wind speed: 11.1 mph

Heating need: Heating is required from September through June and occasionally on August nights, with 56% of the total heating need occurring in December, January, and February. Normal low in January is 11°F, with extreme lows to −24°F.

In a thermally efficient house, solar heating can provide 91% of the total heating requirement, with the solar contribution ranging from 85% in January to 100% April through November.

Cooling need: Relative humidity and temperature are above the comfort range 25% to 33% of the time in July and August, with humidity averaging 56% to 58% and temperature averaging a high of 83°F to 85°F. Extreme high temperatures to 105°F.

In a thermally efficient house, cooling needs can be satisfied naturally by a combination of shading from summer sun and winds admitted to the building at a speed of 3.5 miles per hour.

Detroit, Michigan

Latitude: 42°20′ N
Sun hours/year: 2,375
Degree Days/year: 6,228
Avg. wind speed: 11.7 mph

Heating need: Heating is required from August through June, with 54% of the total heating need occurring in December, January, and February. Normal low in January is 19°F, with extreme lows to −6°F.

In a thermally efficient house, solar heating can provide 94% of the total heating requirement, with the solar contribution ranging from 89% in December to 100% April through November.

Cooling need: Relative humidity and temperature are above the comfort range for short periods in July, with humidity averaging 51% and temperature averaging a high of 83°F. Extreme high temperatures to 105°F.

In a thermally efficient house, cooling needs can be satisfied naturally by a combination of shading from summer sun and winds admitted to the building at a speed of 1.5 miles per hour.

Dodge City, Kansas

Latitude: 37°46′ N
Sun hours/year: 3,219
Degree Days/year: 5,046
Avg. wind speed: 13.2 mph

Heating need: Heating is required from September through June, with 57% of the total heating need occurring in December, January, and February. Normal low in January is 19°F, with extreme lows to −5°F.

In a thermally efficient house, solar heating can provide 99% of the total heating requirement, with the solar contribution ranging from 98% in December to 100% February through November.

Cooling need: Relative humidity and temperature are above the comfort range about 50% of the time in June, July, and August, with humidity averaging 43% to 46% in July and August, and temperature averaging a high of 91°F. Extreme high temperatures to 109°F.

In a thermally efficient house, cooling needs can be satisfied naturally by a combination of shading from summer sun and winds admitted to the building at a speed of 6 miles per hour. However, winds at this speed may cause discomfort to residents simply because of the rapidity of air movement.

Duluth, Minnesota

Latitude: 46°50′ N
Sun hours/year: 2,475
Degree Days/year: 9,756
Avg. wind speed: 11.4 mph

Heating need: Heating is required 12 months of the year, with 49% of the total heating need occurring in December, January, and February. Normal low in January is −0.6°F, with extreme lows to −39°F.

*In a thermally efficient house, solar heating can provide 90%
of the total heating requirement,* with the solar contribution
ranging from 79% in December to 100% June through October.

Cooling need: At no time are temperature and relative humidity
above the comfort range.

*In a thermally efficient house, cooling needs can be satisfied
naturally* with shading from summer sun.

Fargo, North Dakota

Latitude: 46°54′ N
Sun hours/year: 2,586
Degree Days/year: 9,271
Avg. wind speed: 12.7 mph

Heating need: Heating is required 12 months of the year, with
54% of the total heating need occurring in December, January,
and February. Normal low in January is −3.6°F, with extreme
lows to −35°F.

*In a thermally efficient house, solar heating can provide 91%
of the total heating requirement,* with the solar contribution
ranging from 83% in December to 100% May through October.

Cooling need: Relative humidity and temperature are above the
comfort range briefly in July and August, with humidity averaging
55% and temperature averaging a high of 82°F. Extreme high
temperatures to 106°F.

*In a thermally efficient house, cooling needs can be satisfied
naturally* by a combination of shading from summer sun and
winds admitted to the building at a speed of 1.5 miles per hour.

Green Bay, Wisconsin

Latitude: 44°39′ N
Sun hours/year: 2,388
Degree Days/year: 8,098
Avg. wind speed: 10.2 mph

Heating need: Heating is required 12 months of the year, with
52% of the total heating need occurring in December, January,
and February. Normal low in January is 0°F, with extreme lows to
−31°F.

*In a thermally efficient house, solar heating can provide 91%
of the total heating requirement,* with the solar contribution
ranging from 83% in December to 100% May through October.

Cooling need: Relative humidity and temperature are above the
comfort range briefly in July and August, with humidity averaging
57% and temperature averaging a high of 80°F. Extreme high
temperatures to 99°F.

*In a thermally efficient house, cooling needs can be satisfied
naturally* by a combination of shading from summer sun and
winds admitted to the building at a speed of 1.5
miles per hour.

Harrisburg, Pennsylvania

Latitude: 40°16′ N
Sun hours/year: 2,604
Degree Days/year: 5,224
Avg. wind speed: 6.3 mph

Heating need: Heating is required 12 months of the year, with 57% of the total heating need occurring in December, January, and February. Normal low in January is 22.5°F, with extreme lows to −8°F.

In a thermally efficient house, solar heating can provide 94% of the total heating requirement, with the solar contribution ranging from 91% in December to 100% April through November.

Cooling need: Relative humidity and temperature are above the comfort range briefly in June and about 50% of the time in July and August, with humidity averaging 54% and temperature averaging a high of 86.8°F. Extreme high temperatures to 107°F.

In a thermally efficient house, cooling needs can be satisfied naturally by a combination of shading from summer sun and winds admitted to the building at a speed of 3.5 miles per hour.

Indianapolis, Indiana

Latitude: 39°44′ N
Sun hours/year: 2,668
Degree Days/year: 5,577
Avg. wind speed: 7 mph

Heating need: Heating is required from September through May, with 56% of the total heating need occurring in December, January, and February. Normal low in January is 19.7°F, with extreme lows to −20°F.

In a thermally efficient house, solar heating can provide 91% of the total heating requirement, with the solar contribution ranging from 86% in December to 100% April through November.

Cooling need: Relative humidity and temperature are above the comfort range briefly in June and over 50% of the time in July and August, with humidity averaging 60% and temperature averaging a high of 85°F. Extreme high temperatures to 104°F.

In a thermally efficient house, cooling needs can be satisfied naturally by a combination of shading from summer sun and winds admitted to the building at a speed of 3.5 miles per hour.

International Falls, Minnesota

Latitude: 48°34′ N
Sun hours/year: 2,500
Degree Days/year: 10,547
Avg. wind speed: 9.2 mph

Heating need: Heating is required 12 months of the year, with 50% of the total heating need occurring in December, January, and February. Normal low in January is −9°F, with extreme lows to −46°F.

In a thermally efficient house, solar heating can provide 86% of the total heating requirement, with the solar contribution ranging from 73% in December to 100% June through October.

Cooling need: At no time are temperature and relative humidity above the comfort range.

In a thermally efficient house, cooling needs can be satisfied naturally with shading from summer sun.

Louisville, Kentucky

Latitude: 38°13′ N
Sun hours/year: 2,601
Degree Days/year: 4,640
Avg. wind speed: 7 mph

Heating need: Heating is required from September through May, with 58% of the total heating need occurring in December, January, and February. Normal low in January is 24.5°F, with extreme lows to −20°F.

In a thermally efficient house, solar heating can provide 97% of the total heating requirement, with the solar contribution ranging from 94% in January to 100% March through November.

Cooling need: Relative humidity and temperature are above the comfort range part of the time June through September, with humidity averaging 58% and temperature averaging a high of 87.3°F. Extreme high temperatures to 105°F.

In a thermally efficient house, cooling needs can be satisfied naturally by a combination of shading from summer sun and winds admitted to the building at a speed of 4.6 miles per hour. However, winds at this speed may cause discomfort to residents simply because of the rapidity of air movement.

Madison, Wisconsin

Latitude: 43°08′ N
Sun hours/year: 2,502
Degree Days/year: 7,730
Avg. wind speed: 9.9 mph

Heating need: Heating is required 12 months of the year, with 53% of the total heating need occurring in December, January, and February. Normal low in January is 0.8°F, with extreme lows to −37°F.

In a thermally efficient house, solar heating can provide 90% of the total heating requirement, with the solar contribution ranging from 81% in December to 100% April through October.

Cooling need: Relative humidity and temperature are above the comfort range for brief periods in July and August, with humidity averaging 56% and temperature averaging a high of 81°F. Extreme high temperatures to 104°F.

In a thermally efficient house, cooling needs can be satisfied naturally by a combination of shading from summer sun and winds admitted to the building at a speed of 1.5 miles per hour.

Minneapolis and St. Paul, Minnesota

Latitude: 44°53′ N
Sun hours/year: 2,607
Degree Days/year: 8,159
Avg. wind speed: 10.5 mph

Heating need: Heating is required 12 months of the year, with 54% of the total heating need occurring in December, January, and February. Normal low in January is 3°F, with extreme lows to −34°F.

In a thermally efficient house, solar heating can provide 90% of the total heating requirement, with the solar contribution ranging from 81% in December to 100% April through October.

Cooling need: Relative humidity and temperature are above the comfort range for short periods in July and August, with humidity averaging 55% in July and temperature averaging a high of 82°F. Extreme high temperatures to 104°F.

In a thermally efficient house, cooling needs can be satisfied naturally by a combination of shading from summer sun and winds admitted to the building at a speed of 1.5 miles per hour.

New York, New York

Latitude: 40°46′ N
Sun hours/year: 2,677
Degree Days/year: 4,909
Avg. wind speed: 10 mph

Heating need: Heating is required from September through May, with 57% of the total heating need occurring in December, January, and February. Normal low in January is 26°F, with extreme lows to −2°F.

In a thermally efficient house, solar heating can provide 94% of the total heating requirement, with the solar contribution ranging from 91% in January to 100% April through November.

Cooling need: Relative humidity and temperature are above the comfort range for nearly 50% of July and all of August, with humidity averaging 54% and temperature averaging a high of 84°F. Extreme high temperatures to 107°F.

In a thermally efficient house, cooling needs can be satisfied naturally by a combination of shading from summer sun and winds admitted to the building at a speed of 2.3 miles per hour.

Omaha, Nebraska

Latitude: 41°19′ N
Sun hours/year: 2,997
Degree Days/year: 6,049
Avg. wind speed: 10.9 mph

Heating need: Heating is required from August through June, with 58% of the total heating need occurring in December, January, and February. Normal low in January is 12°F, with extreme lows to −22°F.

In a thermally efficient house, solar heating can provide 91% *of the total heating requirement,* with the solar contribution ranging from 83% in January to 100% April through October.

Cooling need: Relative humidity and temperature are above the comfort range much of the summer, with humidity averaging 56% and temperature averaging a high of 89°F. Extreme high temperatures to 114°F.

In a thermally efficient house, cooling needs can be satisfied naturally by a combination of shading from summer sun and winds admitted to the building at a speed of 4.6 miles per hour. However, winds at this speed may cause discomfort to residents simply because of the rapidity of air movement.

Portland, Oregon

Latitude: 45°36′ N
Sun hours/year: 2,122
Degree Days/year: 4,792
Avg. wind speed: 7 mph

Heating need: Heating is required 12 months of the year, with 47% of the total heating need occurring in December, January, and February. Normal low in January is 32.5°F, with extreme lows to −3°F.

In a thermally efficient house, solar heating can provide 96% *of the total heating requirement,* with the solar contribution ranging from 93% in January to 100% February through November.

Cooling need: Relative humidity and temperature are above the comfort range very briefly in July and August, with humidity averaging 65% and temperature averaging a high of 79°F. Extreme high temperatures to 107°F.

In a thermally efficient house, cooling needs can be satisfied naturally by a combination of shading from summer sun and winds admitted to the building at a speed of 1.5 miles per hour.

Rapid City, South Dakota

Latitude: 44°03′ N
Sun hours/year: 2,858
Degree Days/year: 7,324
Avg. wind speed: 10 mph

Heating need: Heating is required 12 months of the year, with 49% of the total heating need occurring in December, January, and February. Normal low in January is 9.6°F, with extreme lows to −27°F.

In a thermally efficient house, solar heating can provide 92% *of the total heating requirement,* with the solar contribution ranging from 85% in January to 100% April through October.

Cooling need: Relative humidity and temperature are above the comfort range very briefly in July and August, with humidity averaging 45% and temperature averaging a high of 86°F. Extreme high temperatures to 110°F.

In a thermally efficient house, cooling needs can be satisfied naturally by a combination of shading from summer sun and winds admitted to the building at a speed of 2.3 miles per hour.

Reno, Nevada
Latitude: 39°30′ N
Sun hours/year: 3,483
Degree Days/year: 6,022
Avg. wind speed: 6.7 mph

Heating need: Heating is required 12 months of the year, with 46% of the total heating need occurring in December, January, and February. Normal low in January is 18.3°F, with extreme lows to −16°F.

In a thermally efficient house, solar heating can provide 99% of the total heating requirement, with the solar contribution ranging from 98% in December to 100% February through November.

Cooling need: Relative humidity and temperature are above the comfort range part of the time in July and August, with humidity averaging 31% and temperature averaging a high of 91°F. Extreme high temperatures to 104°F.

In a thermally efficient house, some of the cooling needs can be satisfied naturally by a combination of shading from summer sun and winds admitted to the building for ventilation. However, these will not be sufficient for cooling in parts of July.

Richmond, Virginia
Latitude: 37°30′ N
Sun hours/year: 2,663
Degree Days/year: 3,939
Avg. wind speed: 6.6 mph

Heating need: Heating is required from September through May, with 60% of the total heating need occurring in December, January, and February. Normal low in January is 27°F, with extreme lows to −12°F.

In a thermally efficient house, solar heating can provide 99% of the total heating requirement, with the solar contribution ranging from 99% in December to 100% February through November.

Cooling need: Relative humidity and temperature are above the comfort range part of the time in June and for much of July and August, with humidity averaging 56% and temperature averaging a high of 88°F. Extreme high temperatures to 105°F.

In a thermally efficient house, cooling needs can be satisfied naturally by a combination of shading from summer sun and winds admitted to the building at a speed of 4.6 miles per hour. However, winds at this speed may cause discomfort to residents simply because of the rapidity of air movement.

St. Louis, Missouri

Latitude: 38°45′ N
Sun hours/year: 2,694
Degree Days/year: 4,759
Avg. wind speed: 9.5 mph

Heating need: Heating is required from September through May, with 59% of the total heating need occurring in December, January, and February. Normal low in January is 22.6°F, with extreme lows to −14°F.

In a thermally efficient house, solar heating can provide 97% of the total heating requirement, with the solar contribution ranging from 95% in January to 100% March through November.

Cooling need: Relative humidity and temperature are above the comfort range for much of the summer, with humidity averaging 57% and temperature averaging a high of 84°F. Extreme high temperatures to 106°F.

In a thermally efficient house, cooling needs can be satisfied naturally by a combination of shading from summer sun and winds admitted to the building at a speed of 4.6 miles per hour. However, winds at this speed may cause discomfort to residents simply because of the rapidity of air movement.

Salt Lake City, Utah

Latitude: 40°46′ N
Sun hours/year: 3,059
Degree Days/year: 5,983
Avg. wind speed: 9.4 mph

Heating need: Heating is required from August through June, with 52% of the total heating need occurring in December, January, and February. Normal low in January is 18.5°F, with extreme lows to −30°F.

In a thermally efficient house, solar heating can provide 96% of the total heating requirement, with the solar contribution ranging from 93% in January to 100% February through November.

Cooling need: Relative humidity and temperature are above the comfort range part of the time in July and August, with humidity averaging 29% and temperature averaging a high of 92.8°F. Extreme high temperatures to 107°F.

In a thermally efficient house, some of the cooling needs can be satisfied naturally by a combination of shading from summer sun and winds admitted to the building for ventilation. However, these will not be sufficient for cooling in parts of August.

Seattle, Washington

Latitude: 47°27′ N
Sun hours/year: 1,783
Degree Days/year: 4,727
Avg. wind speed: 9.3 mph

Heating need: Heating is required 12 months of the year, with 44% of the total heating need occurring in December, January, and February. Normal low in January is 34.7°F, with extreme lows to 11°F.

In a thermally efficient house, solar heating can provide 94% of the total heating requirement, with the solar contribution ranging from 89% in January to 100% February through November.

Cooling need: At no time are temperature and relative humidity above the comfort range.

In a thermally efficient house, cooling needs can be satisfied naturally with shading from summer sun.

Spokane, Washington

Latitude: 47°40′ N
Sun hours/year: 2,019
Degree Days/year: 6,835
Avg. wind speed: 8 mph

Heating need: Heating is required 12 months of the year, with 47% of the total heating need occurring in December, January, and February. Normal low in January is 19.6°F, with extreme lows to −25°F.

In a thermally efficient house, solar heating can provide 90% of the total heating requirement, with the solar contribution ranging from 80% in December to 100% March through November.

Cooling need: Relative humidity and temperature are above the comfort range briefly in July, with humidity averaging 39% and temperature averaging a high of 84°F. Extreme high temperatures to 108°F.

In a thermally efficient house, cooling needs can be satisfied naturally by a combination of shading from summer sun and winds admitted to the building at a speed of 1.5 miles per hour.

Springfield, Missouri

Latitude: 37°14′ N
Sun hours/year: 2,820
Degree Days/year: 4,570
Avg. wind speed: 11.2 mph

Heating need: Heating is required from August through June, with 59% of the total heating need occurring in December, January, and February. Normal low in January is 22.6°F, with extreme lows to −12°F.

In a thermally efficient house, solar heating can provide 98% of the total heating requirement, with the solar contribution ranging from 97% in January to 100% February through November.

Cooling need: Relative humidity and temperature are above the comfort range for much of the summer, with humidity averaging 55% and temperature averaging a high of 89°F in July. Extreme high temperatures to 113°F.

In a thermally efficient house, cooling needs can be satisfied naturally by a combination of shading from summer sun and winds admitted to the building at a speed of 4.6 miles per hour. However, winds at this speed may cause discomfort to residents simply because of the rapidity of air movement.

Syracuse, New York
Latitude: 43°07′ N
Sun hours/year: 2,241
Degree Days/year: 6,678
Avg. wind speed: 9.8 mph

Heating need: Heating is required 12 months of the year, with 53% of the total heating need occurring in December, January, and February. Normal low in January is 15.8°F, with extreme lows to −26°F.

In a thermally efficient house, solar heating can provide 90% of the total heating requirement, with the solar contribution ranging from 83% in December to 100% April through November.

Cooling need: At no time are temperature and relative humidity above the comfort range.

In a thermally efficient house, cooling needs can be satisfied naturally with shading from summer sun.

Youngstown, Ohio
Latitude: 41°16′ N
Sun hours/year: 2,574
Degree Days/year: 6,426
Avg. wind speed: 9.9 mph

Heating need: Heating is required 12 months of the year, with 53% of the total heating need occurring in December, January, and February. Normal low in January is 18°F, with extreme lows to −18°F.

In a thermally efficient house, solar heating can provide 92% of the total heating requirement, with the solar contribution ranging from 86% in December to 100% April through November.

Cooling need: Relative humidity and temperature are above the comfort range for part of the summer, with humidity averaging 56% and temperature averaging a high of 81°F. Extreme high temperatures to 100°F.

In a thermally efficient house, cooling needs can be satisfied naturally by a combination of shading from summer sun and winds admitted to the building at a speed of 1.8 miles per hour.

Climate-Responsive Dwellings

2

After studying and recording the climatic conditions at the building site, the next step in designing your solar house is to understand the climatic forces at work within the house itself, so you can create the most thermally efficient structure possible. Although solar heat is free, it also has its limits. Once the sun sets, you have only so much solar heat stored up. For more free heat, you have to wait until the next morning, and if you're lucky it will be a sunny day. If not, you'll just have to wait longer. In the meantime, you must do everything possible to trap that solar heat in your house by protecting it from the harsh environment.

Since this is obviously not the first generation to seek protection from climate by the way its shelters are designed, let's see what earlier generations on this continent did. For the energy-hungry buildings of today are aberrant in the history of shelter design, and only cheap fossil fuels have made this possible.

Historic Response to Climate

Historically, shelter design has been based upon the need to conserve energy. The reason for this was the scarcity and high cost of fuel—both in terms of actual cost and in its demand for human labor to produce it.

In the Midwest, Native Americans built shelters to suit their needs for warmth, cooling, and security. In the Dakotas, the Mandans built earth-covered structures with low entrances. The entrance stratified air, maintaining constant interior temperatures by keeping the coldest air along the floor. The earth layer

An earth berm and double-entry air lock protect this house from air infiltration.

Two early American
dwellings.
Top: The arched-sapling
framework of the Fox tribe,
to be covered with mats or
sheets of bark.
Above: A traditional
Swedish-style log cabin.

insulated the building. In Wisconsin, the Fox tribe made a
framework of arched saplings, covered with layers of birch and
elm bark, earth, and mats of grasses for insulation. In Missouri,
tribes built oval or round buildings covered with grass. An opening
at the top, together with the entrance, provided ventilation so that
cooking and heating fires could be comfortably built inside the
shelter.

Aztalan, a Native American village in central Wisconsin,
contained buildings made of mud and grass, thickly layered in
response to a harsh climate. And over 10,000 years ago, at Silver
Mound in north central Wisconsin, Paleo-Indians quarried stone
for tools, hunted, and raised crops. No one knows what their
shelters looked like, but it is known that they, like many other
ancient tribes, were able to protect themselves and to survive in a
rigorous winter climate.

The earliest European settlers in the Midwest dug into the
earth, making huts with earthen floors and back walls. The sides,
front walls, and roof were made of wood. These huts modified the
outside climate enough to bring the inhabitants through severe
winters until more permanent homes could be built. For
permanent homes, settlers who lived where wood was plentiful
frequently built log houses, based upon a traditional style brought
to this country by Swedish settlers.

Early uses of native building materials had little to do with
energy conservation, but were based upon the need to make do
with what was at hand—the lack of means to move other
materials to the building site. In the Plains states, buildings were
often made of sod because of this need to build with available
materials. Where clay deposits were nearby, and along rivers
where transportation was feasible, homes were built of brick. In
areas where there was suitable stone, this was used. When wood
was available, however, it was selected over other materials
because of its insulating properties.

As transportation capability increased with the building of the
railroads, wood taken from the forests of the northland became
plentiful throughout the Midwest. Gradually, the predominant
midwestern home became a frame building of stud construction,
with wood siding and wood roof. The typical house had small
windows, slightly raised front and rear porches that were
sometimes screened for summer use, belvederes or vents, and
high roofs to induce natural ventilation. It was compact, two-story,
with cellars for food storage. There were always mud rooms or
vestibules to trap cool air and keep it out of the house. Parlors

Right: A typical midwestern farmhouse (ca. 1900), with small windows and raised porches.
Below: This Italianate-style house (ca. 1865) incorporates a prominent belvedere to induce natural ventilation.

The root cellar of this Federal-style home (ca. 1830) is partially covered with earth and maintains even temperature and humidity to prolong the life of stored vegetables and fruits.

were heated only when company came—important company! Usually, the second floor was not heated, except by gravity. Instead, the sleeping body was warmed with comforters and quilts. Simplicity and common sense were the keys to these carpenter-built homes.

In towns and cities, buildings often shared common side walls to reduce heat loss. Chimneys, for the wood and coal stoves or furnaces, came right through the center of rooms, giving off much of their heat on the way up and out.

As it is today, the problem of the fireplace was that it gave up most of its heat to the out-of-doors, especially when placed on an outside wall. Many times, the builder tried to solve this problem by placing the fireplace in the middle of the building, to reclaim some of the heat given off by the stones or brick. Nevertheless, the fireplace, then as now, still sent much of its heat up the chimney and out. While the fire did warm occupants when they were close to it, it also tended to cool the rest of the house as it pulled interior air to the fire in the combustion process and then up the chimney. The resulting drafts sucked air into the house from the colder outdoors through cracks in the walls.

For insulation, the early builders devised some ingenious methods. In some cases, walls were back-plastered behind the lath-and-plaster finished wall, to provide two layers of plaster for insulation. In our travels through the Midwest, we have seen buildings where solid timbers were set in mortar between the studs for insulative effect. When newspapers were available they served a useful second purpose, as insulation between the studs.

Two examples of the "first generation" of solar architecture in America. Above: The south-facing, solar-heated Lake County Tuberculosis Sanitarium, designed by William A. Ganster in 1940. Below: The Weix House, Oconomowoc, Wisconsin, designed by Keck and Keck in 1960, with south-facing glass, vertical ventilation louvers, wide overhangs, and long east-west orientation.

Some of the best midwestern examples of climate responsiveness in buildings occur, not in houses, but in buildings designed for farm animals. Barns and chicken coops faced south, with high shed-type roofs and south-facing clerestory windows. Large hay mows or piles allowed hay to breathe, preventing heat buildup that could cause spontaneous combustion. Slatted corn cribs naturally dried corn.

For energy sources, settlers turned to wind and water. Mills and pumps powered by these sources performed much necessary work, such as grinding grain and pumping water. Food storage also utilized natural energy. Spring houses chilled milk and kept other perishables safely cool. Root cellars maintained even temperatures and humidity to prolong the life of root vegetables and fruits that rounded out the diet. Even the preservation of food required little energy. Food was dried, pickled, and preserved. Jam, jelly, and sauerkraut stored the summer's bounty.

In the first half of the 20th century, many buildings that were climate responsive in a new way began to appear in the Midwest, designed by midwestern architects. The famous Illinois and Wisconsin architect Frank Lloyd Wright built homes that faced the sun and were directly heated by it. A good example of these is the Jacobs House, built in 1948 near Madison, Wisconsin. William and George Fred Keck of Chicago were pioneers in the design of direct-gain solar houses, beginning with the House of Tomorrow, which they designed and built for the 1933 Century of Progress Exposition in Chicago. The firm of Keck and Keck designed individual houses and also planned whole subdivisions of solar-heated houses. In 1940 a tuberculosis sanitarium that was heated by the direct rays of the sun was built near Waukegan, Illinois. The architect was William A. Ganster, who had a long and

The Von Zellen House,
DeKalb, Illinois.
This page: The north side
has no windows except
at the door.
Opposite page: The
house is oriented so that
all living spaces are
on the south side,
facing the sun.

distinguished career in Waukegan. These early buildings of Wright, Keck and Keck, and Ganster all had wide expanses of glass, broad overhangs, and tile or concrete floors to absorb the sun's energy and radiate it back to the space at night. They belong to the "first generation" of solar houses built in this century.

There are many other examples of climate-responsive shelters, homes, and buildings. But during the last four decades, as most of you already know, aesthetics have played a stronger role in architecture than energy conservation, primarily due to the availability of cheap energy. This period did see the rise of glass-fiber insulation and insulated windows, yet little was done to maximize the use of these conservation measures. There was no need to; beyond a certain point it was deemed a waste of money.

Today, however, climate-responsive buildings are gaining in recognition everywhere, for the same reasons that earlier builders found important. A low energy demand means a considerable savings for the homeowner—in cash, in labor, and in time spent to provide for this part of the household budget.

Achieving Thermal Efficiency

Keeping in mind the lessons learned from past builders, let us look at techniques for building a thermally efficient house today. We are not advocating a return to earlier types of houses. Life-styles have changed, the cost and availability of energy have changed, and technology and materials have changed. Therefore, a new and improved kind of climate-responsive house can be built—the passive solar house. It is designed not only to use

The Von Zellen House

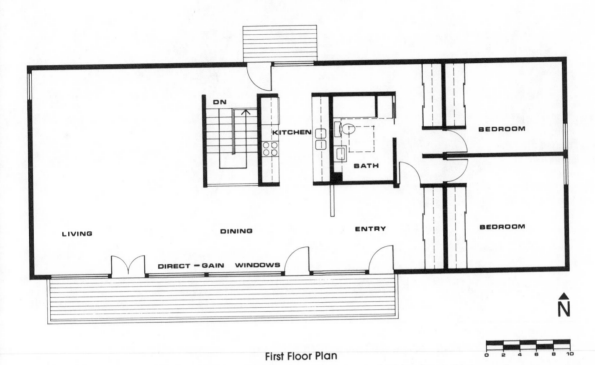

LIVING DINING ENTRY

DN KITCHEN BATH BEDROOM

DIRECT — GAIN WINDOWS

BEDROOM

N̂

0 2 4 6 8 10

First Floor Plan

The Jablonski House, Wauconda, Illinois, is integrated into a south-facing hill, as seen from the east (above) and the southwest (right).

natural energy for heating and cooling, but also to be protected from harsh climate conditions that would steal heat from the house in cold weather or overheat it in summer. That is, it must be designed to be thermally efficient.

To begin, you should orient the house, or set it on the site, so that all of the living spaces face the sun. If all of these spaces—living room, bedrooms, and so on—face south, they can then be warmed directly by the sun. Second, you should protect the house from cold winter winds out of the north and west. Place your garage on the west side of the house and put all of the utility and storage space on the north side. This creates a buffer between prevailing winds and the sun-warmed living spaces on the south side of the building. Since more heat is lost through windows than through walls, try to avoid windows on the cold north side of the building. Windbreaks at the west and north will give you even more shelter from the wind's force. Plant evergreen trees, at least three or four rows deep, to provide a dense stand of trees that will deflect the wind around and over the house.

An effective way to keep cold air from entering your house is earth integration, which means building the house partially or

The Jablonski House

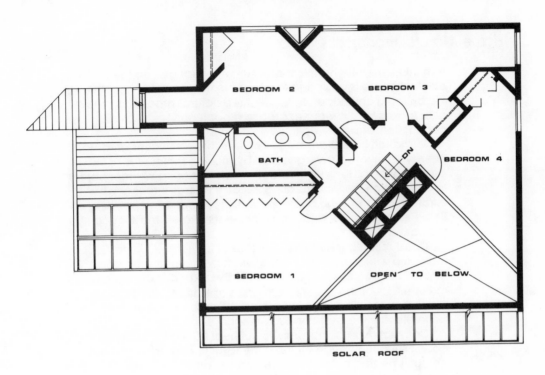

BEDROOM 2

BEDROOM 3

BATH

DN

BEDROOM 4

BEDROOM 1

OPEN TO BELOW

SOLAR ROOF

Second Floor Plan

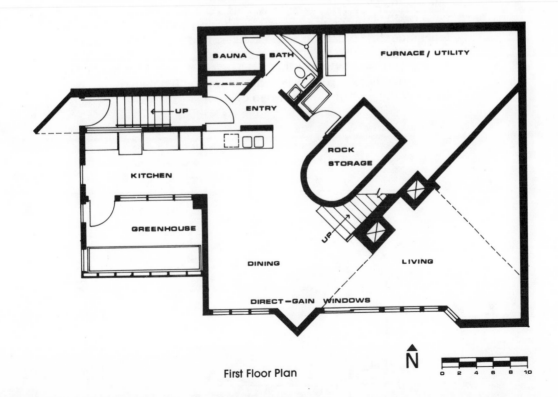

SAUNA BATH

FURNACE / UTILITY

ENTRY

UP

ROCK STORAGE

KITCHEN

UP

GREENHOUSE

DINING

LIVING

DIRECT-GAIN WINDOWS

N

0 2 4 6 8 10

First Floor Plan

Heating Your Home: Natural Processes

To minimize winter heat loss and summer heat gain, a thermally efficient house is designed to inhibit the movement of heat through the building envelope. There are three natural processes by which heat moves from a warmer area to a cooler one: radiation, conduction, and convection.

Radiation is the transmission of heat through waves of radiant energy, or infrared rays, given off by moving molecules at the surface of an object. Heat radiation travels from a warmer object to a cooler one without heating the air it passes through. When the infrared rays strike the cooler object, they heat it by speeding up its molecules. It is by radiation that the sun warms the surface of the earth.

Conduction is the movement of heat from molecule to molecule within an object, or between two touching objects. Heated molecules vibrate more rapidly and strike adjoining molecules, which vibrate in turn and strike others. This process distributes heat throughout the object. Dense materials, such as concrete, are the best conductors of heat.

The natural processes of radiation, conduction, and convection distribute heat within a passive solar house.

Convection is the transmission of heat from a warmer to a cooler surface through air movement. Air in contact with a hot object is heated by conduction and expands and rises, distributing its heat throughout the room. When it has given off its heat it becomes denser and sinks, to be heated once more when it contacts the hot object. The flow of heated air away from a hot object and cooler air toward it is called a convection current.

Chapter 3 discusses passive solar design and construction techniques that make the house itself the heating system, taking advantage of these natural processes to collect the sun's energy, store it, and use it to heat a thermally efficient space. The same techniques that keep a house warm in winter help keep it comfortably cool in summer when outdoor temperatures are high.

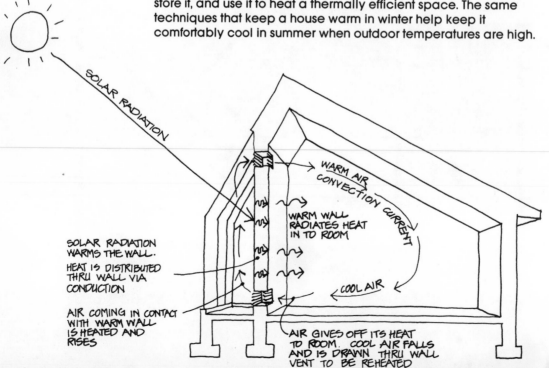

SOLAR RADIATION

WARM AIR CONVECTION CURRENT

WARM WALL RADIATES HEAT IN TO ROOM

SOLAR RADIATION WARMS THE WALL.

HEAT IS DISTRIBUTED THRU WALL VIA CONDUCTION

AIR COMING IN CONTACT WITH WARM WALL IS HEATED AND RISES

COOL AIR

AIR GIVES OFF ITS HEAT TO ROOM. COOL AIR FALLS AND IS DRAWN THRU WALL VENT TO BE REHEATED

completely underground, except at the south end. You can build the house into a south-facing slope or you can place earth against the walls of a house that is built above ground and designed to have these earth berms, or mounds, pushed against it. In either case, the house will be open to the sun on the south side, while the other walls will be earth-covered. An earth-sheltered house can be cozy and comfortable, light and airy, with wall hangings and paintings, plants, and brightly colored fabrics to make the space cheerful and light, and with south-facing windows to welcome the sun.

When you build an underground house, be careful to waterproof and to insulate correctly. Several waterproofing systems are available, varying in cost. The waterproofing material must be able to withstand freeze-thaw and moisture conditions. Roofing materials will not work, because they will rot and will need to be replaced.

Bentonite panels are one effective system. Bentonite is a natural clay product, mined in the Dakotas. It is available in 4-by-8-foot sheets, coated with paper fiber. The panels are nailed to the surface that requires waterproofing. With exposure to moisture, the nails rust, the fiber coating disintegrates, the clay particles of Bentonite lock together, and an impermeable surface is formed, repelling moisture.

Membrane systems, made of petroleum-based compounds,

Earth integration protects the Pleasant Valley Outdoor Center of Woodstock, Illinois (shown under construction), from infiltration of cold air. The center is built into a south-facing hill.

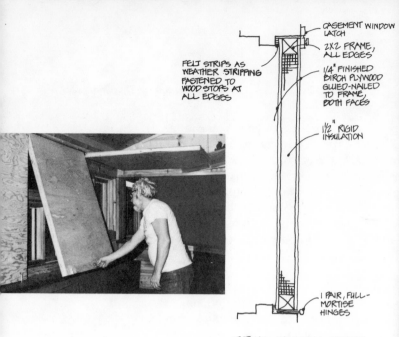

CASEMENT WINDOW
LATCH

2X2 FRAME,
ALL EDGES

1/4" FINISHED
BIRCH PLYWOOD
GLUED-NAILED
TO FRAME,
BOTH FACES

FELT STRIPS AS
WEATHER STRIPPING
FASTENED TO
WOOD STOPS AT
ALL EDGES

1½" RIGID
INSULATION

1 PAIR, FULL-
MORTISE
HINGES

TYPICAL INSULATING SHUTTER

Thermal shutters, filled with rigid insulation, are designed to fit tightly over windows at night to prevent heat loss. They can be installed in a number of ways.

completely wrap the surfaces that are exposed to earth. Somewhat more expensive than Bentonite, a membrane system, when properly applied, will form a coating that is inviolate. As with the Bentonite clay, moisture will be unable to flow through the waterproofing. Any breach of the membrane will permit movement of moisture, so it is important that application be carefully performed and that nothing be permitted to puncture the material. The Bentonite material, if punctured, will heal itself.

Three inches of rigid insulation should cover the walls on the outside, over the waterproofing. Earth will be placed against the insulation. The floor and roof of the earth-integrated house will be made of concrete, and walls will be either concrete or masonry.

Once you have provided for protection from winter cold, the next strategy in designing a thermally efficient house is to counteract summer heat. Windows should be placed so that they will permit free flow of air in summer for its cooling effect. This means placing some windows where they will catch prevailing breezes and placing others higher up, on the leeward side of the house, where they will naturally pull out this breeze. Two other essential cooling elements are shade trees and overhangs: Both should be planned to block the hot summer sun from the house. The details of how we plan these windows and overhangs are covered in the next chapter, under the Cooling section.

To summarize, your goal in designing a thermally efficient

The Sandy Pines Convenience Center of Grand Rapids, Michigan, uses earth berms on all sides of the building to keep the interior cool in summer and warm in winter.

house is to admit winter sun and summer winds and to shut out winter winds and summer sun. The strategies we have described so far are your first line of defense. Your second line of defense against climatic conditions is how you build the house. The house must be tightly built to keep out winds and drafts. It also must be heavily insulated to keep in heat. You want to minimize heat loss, so that less energy will be required to heat your house. The sun won't shine longer to compensate for poor construction, so the house must be adapted to its climate and to the sun.

Generally, windows should be triple layered, to provide more protection from cold outside air. However, on the south side of the house double glazing is best, because the gain in solar radiation through the windows is greater with only two layers of glass. Thermal shutters, shades, or drapes, tightly fitted to the windows, reduce nighttime loss of heat through the windows.

The shutters we use are made of a 1½-inch wood frame with a core of 1½ inches of rigid insulation covered with ¼ inch of finished plywood. They are made to fit tightly in window openings and are weather-stripped so that heated air from the house cannot circulate against the cold window glass. Painted or covered with wallpaper, they become a decorative element for the room. They are hinged to open into the room and are closed at night to hold heat in the house. These shutters are highly effective in reducing loss of heat from the house.

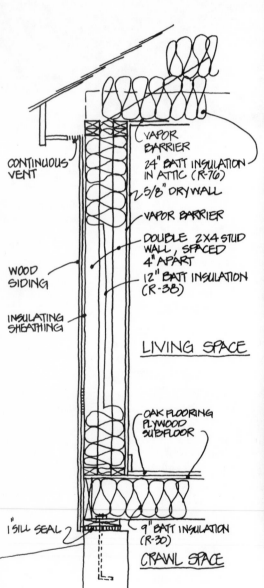

CONTINUOUS VENT

WOOD SIDING

INSULATING SHEATHING

VAPOR BARRIER

24" BATT INSULATION IN ATTIC (R-76)

2 5/8" DRYWALL

VAPOR BARRIER

DOUBLE 2×4 STUD WALL, SPACED 4" APART

12" BATT INSULATION (R-38)

LIVING SPACE

OAK FLOORING PLYWOOD SUBFLOOR

1" SILL SEAL

9" BATT INSULATION (R-30)

CRAWL SPACE

All outside doors should be built as old-fashioned vestibules were, with an outer hall protected by two doors—one to the outdoors and one to the house. This air lock will reduce heat loss as you enter and leave the house. Of course, children can leave two doors open as easily as one, but new habits will develop with time.

For further defense against climate, all openings for windows, doors, and vents should be caulked and weather-stripped. At every break in the envelope of the house that might permit leaking of heat—all cracks and holes—you must seal the openings. Even where the wall joins the floor, a special insulating strip called a sill seal, often 1 inch of fiberglass insulation, is laid over the bolts in the floor before the wood plate is put in place. This insulation will fill any cracks or dips and help to hold heat in the house.

We put 12 inches of insulation in the walls and 24 inches in the ceiling, and we find that the minimal additional cost for this amount of insulation is far outweighed by reducing heat loss. Years ago, when building costs came in above the budget, the contractor always suggested leaving out the insulation, but those days are gone.

To increase the thermal efficiency of the house, always use a vapor barrier, which is a clear plastic film, 4 or 6 mils thick. This should be placed at the inside of the wall or ceiling directly between the insulation and the drywall. Where ceiling and walls meet, the vapor barrier should extend at least a foot over the ceiling to insure an unbroken seal. At floors, place the vapor barrier beneath a concrete slab or over insulation and framing before the subfloor goes down.

Once these requirements are met, you're "halfway home" to your solar house. Now let's look at the systems for providing it with solar heat.

Passive
Solar Systems

3

In order to use the sun to provide the greatest possible portion of heat for your house, it is important to build a house that is thermally efficient, designed in such a manner that the building itself is an effective modifier of local climatic conditions. This means that less energy is required for heating the house. Once you have oriented your house for its environment, you must decide how to develop the building itself. This chapter discusses natural energy solutions for space heating and cooling and also outlines energy-saving means of providing your house with hot water and lighting.

Space-Heating Systems

In meeting your space-heating requirements, you want to select a method or methods of trapping and storing heat from the sun that will provide the desired amount of heat as a portion of the total heating load. In our buildings, we use four passive solar-heating methods: direct gain, the mass heat wall, the sunspace, and the solar roof.

Direct Gain. The simplest and most straightforward method of using the sun is by direct heat gain. The direct-gain system is composed of south-facing, double-glazed windows to collect the sunlight, and large, heavy building components, placed where they will receive the direct rays of the sun, to function as a heat-storage, or thermal, mass. The storage components that we commonly use are a concrete floor slab, an interior masonry wall, or a fireplace. Sunlight enters the house through the windows,

The exterior of a greenhouse (ground floor) and solar roof (second floor).

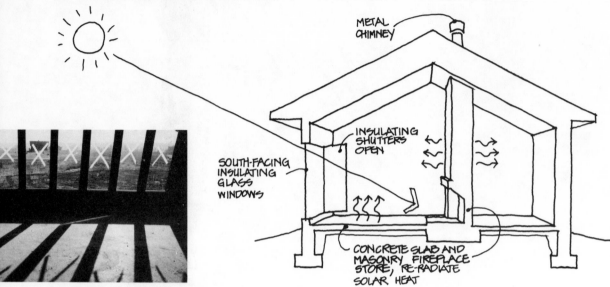

Above, left: Sunlight streams through the newly installed windows of a direct-gain system.
Above, right: How direct-gain solar heating works. The chimney is made of metal above the roof line to prevent the stored heat from being "wicked" off into the cold air.

strikes this thermal mass, and heats it. When the sun is not shining, the mass slowly releases its stored heat to warm the space within. Even when outdoor temperatures are well below 0° Fahrenheit, a thermally efficient house will be comfortably heated. The windows are fitted with insulating shutters, which are closed at night to hold in the heat collected during sunlight hours. Without these shutters, you'd lose more heat at night than you gained during the day.

The square footage of south-facing glass required ranges from 25% to 100% of the square footage of floor space in your thermally efficient house, depending upon local climate and the number of Degree Days. This much glazing will provide 90% to 100% of your space-heating requirements. The amount of thermal mass needed for storage depends on the amount of south-facing window glass you use for direct gain. For every square foot of solar glazing, you will need to provide 200 pounds of mass that the sunlight strikes directly. (Thermal mass not in the direct path of sunlight will also be effective in storing heat. You'll just need more of it, in this case, 600 pounds of mass per square foot of glass.) A cubic foot of concrete, by the way, weighs approximately 144 pounds.

How will the direct-gain house look? It will look very much like other houses that you see around you, except that each living space—living room, kitchen, dining area, and bedrooms—will be directly exposed to the sun and will be heated directly by its own

The kitchen of the Theios House, Wadsworth, Illinois, is warmed directly by the sun. Note the quarry tile flooring, for storage of direct-gain heat.

south-facing windows. Utility spaces—laundry, storage, and bathrooms—will be on the north, protecting living spaces from infiltration of cold air. The storage mass itself, being made of normal building components, will not seem out of place either. The net result is a house that is rectangular in shape, long and narrow along an east-west axis to permit best access to the sun's heat, and built out of standard building materials.

Here is how the direct-gain house is put together. If you use a concrete floor slab as the thermal mass, the slab is normally 4 inches thick with 4 inches of gravel beneath it and rigid insulation underlying the gravel. The slab cannot be covered with carpeting, since this would interfere with the storage process. However,

The Theios House, designed by Rodney Wright in 1961, uses the principle of direct-gain heating, with floor-to-ceiling windows on the south side and wide overhangs to prevent overheating in summer.

throw rugs or a carpet or island rug may be placed where the sun does not strike.

The concrete can be finished in a number of ways, to satisfy your aesthetic sense. The most economical method is to polish it and use it as it is, but you could also have it colored or stained, or even have it scored to resemble brick or tile. Another good heat-storage floor material is quarry tile, which is available in several sizes and a number of configurations and colors—all pleasant to look at. Stone makes a durable floor and adds to the

The Theios House

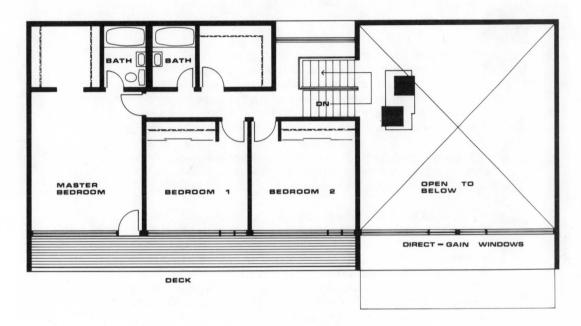

MASTER BEDROOM

BEDROOM 1

BEDROOM 2

BATH

BATH

DN

OPEN TO BELOW

DIRECT - GAIN WINDOWS

DECK

Second Floor Plan

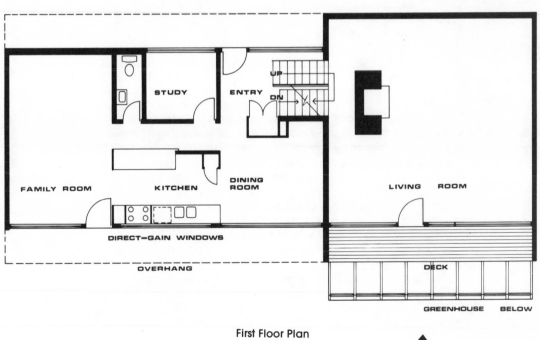

STUDY

ENTRY

UP

DN

FAMILY ROOM

KITCHEN

DINING ROOM

LIVING ROOM

DIRECT—GAIN WINDOWS

OVERHANG

DECK

GREENHOUSE BELOW

First Floor Plan

N

0 2 4 6 8 10

Glazing: Trapping the Sun

Anyone who opens the door of a car that's been left in the sun with its windows rolled up has experienced a basic principle underlying passive solar-heating systems. It is what is often referred to as "the greenhouse effect." In this case, solar radiation passes through the car windows and is absorbed by and heats the interior mass and thus the space within the car. The interior of the car will remain warm for a time after the sun goes down because the heat radiation given off is trapped inside by the windows that let in the sunlight to begin with. The same phenomenon occurs with passive solar-heating systems.

What glazing does. Many passive heat-collection systems begin with a south-facing wall of a transparent or translucent material, usually glass or fiberglass, referred to as "glazing." The glazing admits sunlight into a confined space — a living area, an attached greenhouse, an area of the attic, a narrow air space between the glazing and a wall. When the sunlight strikes a mass within the space, such as concrete floors or masonry walls, the solar radiation is converted to thermal radiation, or heat. Because glazing freely transmits short-wave solar radiation but not long-wave thermal radiation, heat given off by the heavy mass is trapped within the confined space and will not pass through the glazing except via conduction. Applying glazing in two or three layers separated by an air space dramatically reduces this conduction heat transfer.

What kind of glazing should you use? Glazing material can be of several kinds: low-iron-content glass, regular glass, fiberglass (FRP), or fiberglass. Of the three glazing materials we recommend, glass that is low in iron content and fiberglass are the most efficient transmitters of solar radiation; glass with high iron content, the kind most commonly used for everyday purposes, blocks solar radiation to some extent. Fiberglass expands in warm weather and shrinks in cold weather. As a result, it has a taut, trim appearance in cold months, but may "oilcan," or become wavy, and be less attractive in hot months. This cosmetic problem is offset, however, by two factors. The first is cost: Fiberglass costs less than glass. The second factor is that fiberglass is easy to work with. It can be cut with a knife and nailed in place. It is also lighter than glass and easier to handle, and is break resistant.

The choice of glazing material will be influenced by initial cost, installation cost, appearance, and transmissivity. Each of you must weigh these factors and select what is appropriate for your project.

How much glazing do you need? This chapter offers guidelines regarding the amount of solar glazing needed to satisfy the space-heating requirements of a thermally efficient house. Generally speaking, an area of solar collection surface equal to 25% to 100% of the square footage of floor space can supply 90% or more of the space-heating needs. For every square foot of solar glazing, 200 pounds of mass directly heated by the sun for direct gain (or 600 pounds if not directly heated) or 150 pounds of mass heat wall will be needed for heat storage.

In measuring the solar collection surface, care must be taken to subtract the area of glazing that is blocked from direct sunlight by the glazing channels that hold the glazing material in place. An 8-by-8-foot wall of glazing (64 square feet), for example, contains only about 61 square feet of collection surface if the glazing panels are 24 inches wide and are held in place by 1/2-inch-wide channels.

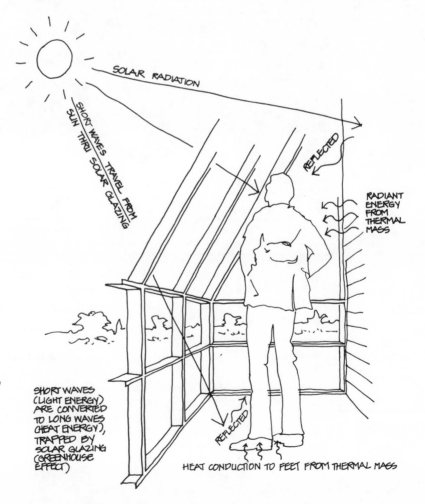

How solar glazing works. Solar radiation passes through glazing and is converted into heat by the thermal mass. A small portion of this heat is conducted by the glazing to the exterior; the remainder is trapped inside and travels in convection currents through the space.

SOLAR RADIATION

SHORT WAVES TRAVEL FROM SUN THRU SOLAR GLAZING

REFLECTED

RADIANT ENERGY FROM THERMAL MASS

SHORT WAVES (LIGHT ENERGY) ARE CONVERTED TO LONG WAVES (HEAT ENERGY), TRAPPED BY SOLAR GLAZING (GREENHOUSE EFFECT)

REFLECTED

HEAT CONDUCTION TO FEET FROM THERMAL MASS

Two details of direct-gain systems.
Above: A wide overhang keeps the Von Zellen House cool in summer.
Right: A stone floor, warmed by direct-gain radiation, acts as a "heat sink," releasing warmth over time to the living space.

storage capacity. Slate, though expensive, is available in elegant shades of black, green, and blue to make a handsome floor.

A wall of 4-inch brick or solid masonry in the direct path of the sun and not more than 16 feet from the south windows will also store heat effectively, but we find this storage mass more expensive to build than other types. If you choose this method of storage, make certain that the sun will strike the wall directly and fully. The masonry itself must be exposed, with no furniture or wall hangings interposed between the wall and sun.

A word of caution, too, about fireplaces used as a storage mass. While the masonry will effectively store heat, you will want to build a fireplace that is energy efficient not only in how it operates, but also in how it is put together. The masonry should stop just below the roof line with an insulated metal chimney continuing above the roof line. Otherwise, heat will travel up through the masonry and be lost to the cold outdoors.

Direct-gain houses may overheat in spring and fall, when less heat is needed but while the sun is fairly low in the sky. Broad roof overhangs, which effectively shade the window area in summer, when the sun is higher in the sky, will partially block the sun at these times. Another way to counteract overheating is simply to open a window! Other, more sophisticated ways of ventilating are discussed later in the book.

If you already have a house, you may be able to retrofit, or remodel, it for direct-gain solar heating. Before undertaking the solar retrofit of a house, you should do everything you can to upgrade it thermally—caulking, weather-stripping, insulating heavily, adding storm windows and thresholds and air locks. Once you have accomplished all of this, and if your house is not obstructed from receiving the sun at the south wall, you can certainly proceed with adaptation of that south wall for direct gain.

Most of our retrofit projects have been done in conjunction with house additions. In these cases, the direct-gain glazing can easily be built into the new addition and planned to serve both the addition and the existing building, as long as it has been made

A solid masonry wall on the interior of the Alan House Motel, Osseo, Wisconsin, lies in the direct path of the sunlight and stores heat during the day for use at night.

thermally efficient. There will be an increase in the overall amount of south-facing glass in the house, because of the glazing in the addition. If they are needed, you can also add new windows to the existing house. A word of caution: If you are installing a window in an existing wall, always remember that you are cutting through a load-bearing wall. In other words, the wall is supporting the weight or load of floors, roof, and/or walls. Over each opening it will be necessary to put a lintel that will carry the load. The wall load and the lintel must be sized by an architect or engineer to be certain that there is no settlement of the building. During construction, you must shore up or support the wall until the lintel is placed.

A word about decorating your direct-gain house might be helpful. Since the building is the heating system, the windows *must* have access to the sun and the thermal mass *must* be uncovered. We have seen homes that have lost their direct-gain capability because the owner carpeted the concrete floor, which was designed to store heat, and then bought drapes to block out the sun so the carpet would not fade. Direct gain is then entirely excluded. Be sure that furnishings exposed to the sunlight are fadeproof—or that you are willing to accept some fading in return for a lot of heat. Decorate your direct-gain house for a life in the sun.

Mass Heat Wall. Also called the Morse Wall, for E. L. Morse from Salem, Massachusetts, who invented it in 1881, or the Trombe Wall, for Felix Trombe, the French physicist who further developed it. This wall gives you, in one configuration, both the collector of solar heat and the storage mass. Built as the south wall of your house, the thermal mass can be made of concrete, solid or filled concrete block, solid brick (yes, some bricks have holes in them), solid stone, or a combination of the above. The exterior face of the wall is covered with three layers of glazing. Between the wall and the glazing is a 3½-inch air space (the width of a normal 2-by-4) in which heat will be trapped when the sun is shining on the wall. With nighttime insulation over the wall, two layers of glazing would be sufficient. However, it is difficult to insulate, so we use triple glazing instead. Triple glazing provides greater protection from nighttime heat loss.

The wall heats your house in two ways. First, the wall soaks up the sun's heat directly, requiring about six hours for the heat that the wall absorbs to move through its 12-inch depth and radiate heat into the house. The heat that the wall is soaking up during the day is thus available for nighttime heating of your house.

Second, the wall creates a flow of sun-heated air that warms the house in the daytime. Through vents in the bottom of the wall, cool house air enters the air space, rises as it is heated, and re-enters the house through the top vents. This process continues naturally, through convective air movement, until the wall cools.

Lightweight backdraft dampers at the bottom vents swing open to let cool air flow from the room into the space between the wall and the glazing, but fall into place to cover the vents when air moves in the opposite direction, preventing cold air from re-entering the house at night. Dampers can also be placed at the top vents, if desired, to be opened manually when the sun has heated the air.

In areas such as the Midwest, where winter temperatures are ordinarily low in the daytime—frequently below freezing—the two methods of taking heat from the wall are a necessity. Without the vent system, there would not be sufficient daytime heat inside the house. In areas of the country where daytime temperatures are warm and heat is needed only at night, vents may not be required.

You will want windows for light and ventilation on the south wall alongside the mass heat wall. According to typical building codes, the minimum in square footage is 10% of the room's area for light with 5% for ventilation—in other words, for a

Below, left: How a mass heat wall works when heating.
Below, right: The glazed exterior of a mass heat wall at the Alan House Motel. Note the direct-gain windows used in conjunction with the mass heat wall. Each motel room has a heating unit as well, although no backup heat was needed in February of 1980, shortly after the building was completed.

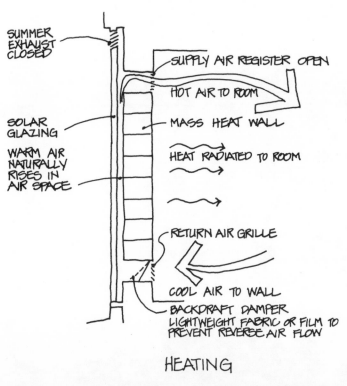

SUMMER EXHAUST CLOSED

SUPPLY AIR REGISTER OPEN

HOT AIR TO ROOM

SOLAR GLAZING

MASS HEAT WALL

WARM AIR NATURALLY RISES IN AIR SPACE

HEAT RADIATED TO ROOM

RETURN AIR GRILLE

COOL AIR TO WALL

BACKDRAFT DAMPER LIGHTWEIGHT FABRIC OR FILM TO PREVENT REVERSE AIR FLOW

HEATING

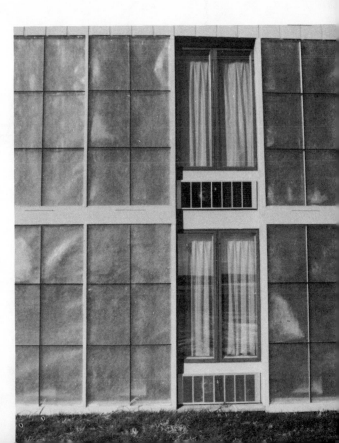

200-square-foot room, you need 20 square feet of glass, half of which will open. The double-glazed windows (with insulating shutters, of course) will provide heat directly, so you will have a combination of mass heat wall and direct gain.

We are delighted and intrigued by the simplicity of the mass heat wall. No one needs to invent new materials or devices. Masons can build the walls easily. Note, however, that we are talking about a 12-inch-thick solid masonry or concrete wall. If hollow block is used, it must be filled with solid concrete. The mason must be made to understand this, since he traditionally fills the cores with paper and then puts concrete at the top only, to hold the anchor bolts. Unless you have used a dark brick or stone, the outside face of the wall should be stained black with a concrete flow stain to better absorb the sunlight.

The vent openings for the mass heat wall are sized in the following manner: For every 24 inches in length of wall, you need an 8-by-8-inch opening at top and bottom. This is the gross size, allowing for the framing to receive a vent. Lintel block with reinforcing rods should be used at these openings.

The Zoes House of Woodstock, Illinois, under construction. The building has a two-story mass heat wall made of black-stained concrete blocks. Solar glazing and a greenhouse on the first floor are yet to be added. The house is built on a south slope, with the driveway/entry on the second floor, allowing the lower floor to be earth-integrated.

The Zoes House

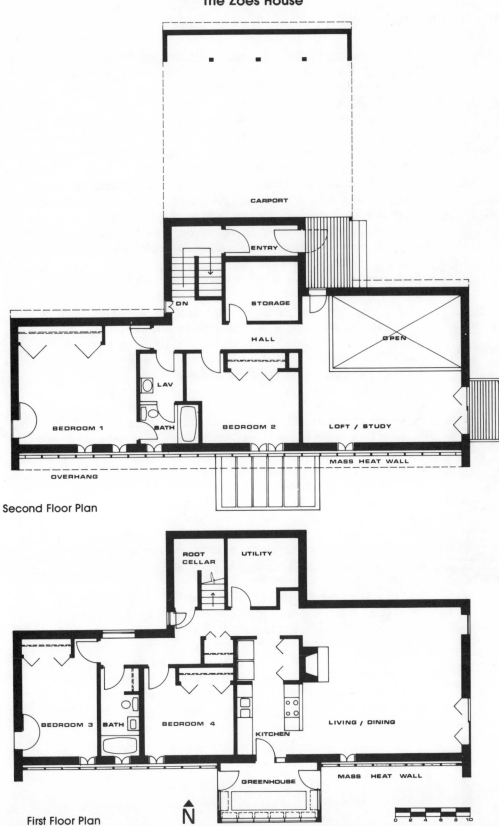

CARPORT

ENTRY

DN

STORAGE

HALL

LAV

OPEN

BEDROOM 1

BATH

BEDROOM 2

LOFT / STUDY

MASS HEAT WALL

OVERHANG

Second Floor Plan

ROOT CELLAR

UTILITY

BEDROOM 3

BATH

BEDROOM 4

KITCHEN

LIVING / DINING

GREENHOUSE

MASS HEAT WALL

First Floor Plan

N

0 2 4 6 8 10

The mass heat wall of the Zoes House under construction. Note how the mason has filled the hollow concrete blocks with additional concrete.

In calculating the size of a mass heat wall, we assume an 8-foot-high wall for a one-story house. For a two-story house, the masonry wall is continuous for the two floors, but has separate pairs of wall vents at each floor.

A new masonry wall, with fresh mortar and water used in the building of it, takes a year to dry out. This is true of all masonry walls and will be true of your mass heat wall. Until the wall is completely dry, it will not perform at peak efficiency.

Quality of workmanship is very important for a successful mass heat wall, as it is for any successful solar building. Everything must be airtight, so that you don't lose heat from the wall to the outdoors. To achieve this airtight fit for our wall, we use one-part acrylic terpolymer sealant in the glazing channels with good results. The glazing channels, or the mullions that receive the glazing, can be wood or metal and are built much like a normal window wall.

In a thermally efficient house, you need a square footage of glazing in a south-facing wall equivalent to between 25% and 100% of the floor space to provide 90% to 100% of the required space heating. There should be 150 pounds of mass wall per square foot of glazing. The glazing material can be low-iron-content glass, regular glass, or panels of glass-fiber-reinforced polymer, or fiberglass. The low-iron-content glass is the most expensive choice and fiberglass panels the least costly.

Now for a word about energy savings in how your new system is built. The energy used to ship materials from place to place should be considered as an "energy cost" of your building. Concrete block is made in all states, but the cement used in making it is not found everywhere—Wisconsin, for example, has none. Similar questions must be asked about the clays for brick. In some rural areas, stone is available for the taking, the result of clearing fields for agriculture. If such material can be obtained easily, perhaps you or a mason can provide a wall at a very reasonable price—to everyone.

There is a potential for overheating with the mass wall in summer. One solution is to install at the top of the heat space small vents fitted with a spring set to open to the out-of-doors at a predetermined temperature (usually in excess of 180°). With these vents open, excess heat will be exhausted to the outside. Because the sun is high in the sky in summer, the vertical wall will receive the least amount of Btu's during this season, and the vents will work nicely. We used this solution at Alan House Motel in central

Wisconsin with good results. Manually operated vents can also be used, with damper handles extending through the wall to the inside of the building.

At times in the summer, when there are cooler temperatures outside and a need for ventilation, you can also use your mass heat wall to your advantage—by making it perform as a fan. If you crack open a window on a shaded wall of your house, open the outside vents on the heat wall, and close the top inside vents, the hot air rising off the wall will be exhausted to the outdoors, pulling cool air from the shaded side into the room. You may still need the stored heat of the wall during cool summer nights, so it will be of help to you in more ways than one—working as a fan during the day and as a heater at night.

The retrofit of an existing house for a mass heat wall is perhaps the most difficult of the solar retrofits—especially if you're dealing with a wood-frame house. If you want to build such a wall for an existing frame house, in most cases a new foundation wall is needed to support the weight and thickness of the mass heat wall. You can try to tear out an entire south wall from a frame house and temporarily shore up the roof, but the job is so extensive and messy and expensive, with such questionable end results, that we prefer to build a new foundation up against the old, and erect a new mass heat wall outside the frame wall. Once

Below, left: How a mass heat wall works when cooling.
Below, right: The owner of the Alan House Motel checks an air-intake vent on a mass heat wall; each room has two cold-air intakes, one hot-air return, and an exterior vent for summertime use.

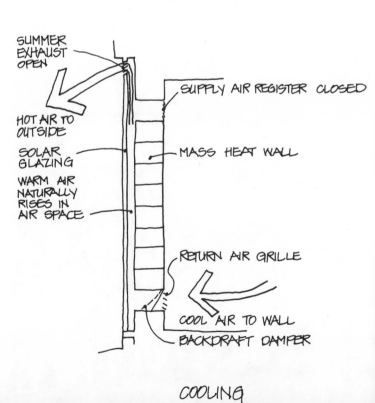

SUMMER EXHAUST OPEN

HOT AIR TO OUTSIDE

SOLAR GLAZING

WARM AIR NATURALLY RISES IN AIR SPACE

SUPPLY AIR REGISTER CLOSED

MASS HEAT WALL

RETURN AIR GRILLE

COOL AIR TO WALL

BACKDRAFT DAMPER

COOLING

the wall is done, the frame wall is removed and the new wall is tied into the house. This is a sizable job that a beginning builder should not tackle alone, but one that can provide an effective heating method nonetheless.

If the house to which you wish to add a mass heat wall is built of masonry, all you have to do is add glazing and cut vents into a south-facing wall, being sure to stain or darken the exterior masonry before you glaze. The finished wall will resemble a window wall when seen from the exterior, and no one in your neighborhood should object to how it looks.

When you decorate the inside of your house, make sure that your mass heat wall is left bare, exposing the block, concrete, or brick to the air space. You can paint the wall to match other wall surfaces inside the house. If you don't like the exposed material, you can plaster with portland cement directly to the block, or paint with a sand-aggregate paint. This does not impair the efficiency of the wall, but it does increase your costs a bit.

We believe that the mass heat wall is ideally suited to many purposes other than housing. Warehouses, garages, factories, schools, offices—any building where you want exterior walls of masonry—can have a south-facing mass heat wall. This is especially desirable in buildings where direct sunlight may not be acceptable, such as schools and offices. Concrete panels that are poured on site and tilted up into position can be cast with the vent openings already in place. This is truly an efficient way of using the sun, since the wall serves both a heating function and a structural function.

Sunspaces. A sunspace is a glazed space along the south wall or roof of a house that traps the heat of the sun and transmits it to the inside of the house. The sunspace can be planned as a

The Anderson House, Woodstock, Illinois. Opposite page: Interior of a sunspace (top). The interior wall of brick acts as a thermal mass to store heat during the day. Sunspaces are located along the length of the south side of the house (bottom). This page: A cardboard model built to scale, and the floor plan.

greenhouse, doing double duty by growing your food and flowers while it heats your house. Or you can build a solar attic at the roof level. The attic is useful in locations where the sun, because of obstructions such as buildings, trees, hills, or mountains, does not strike the floor line.

Let's look at each type of sunspace in turn. First, the plain and unadorned sunspace is narrow, typically 3 or 4 feet in depth, and glazed on its entire south wall. The sun heats this space and heat then flows into the living space of your house. Heat may be transferred by opening up the south wall of the room to the sunspace, or the room wall can be glazed to work like a direct-gain wall. Fans can be used to move the heated air, or you

The Anderson House

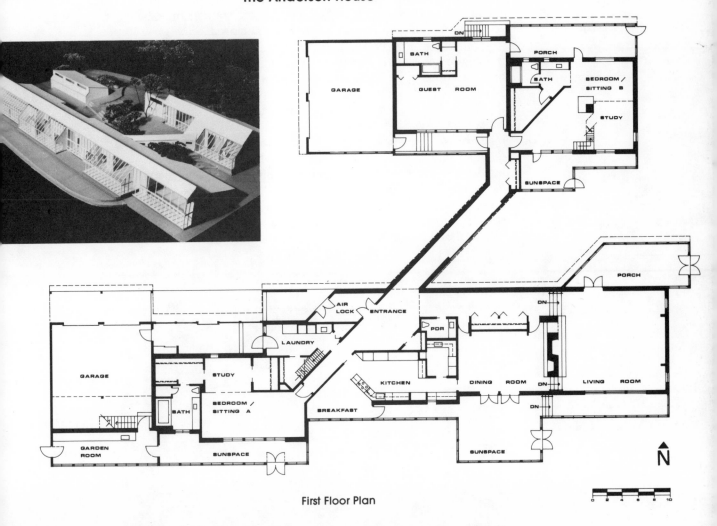

First Floor Plan

can permit it to move naturally, by convection currents, into the cooler space of the house. Insulating shutters between the sunspace and the house are necessary to protect the house from nighttime heat loss in winter and from daytime overheating in summer.

A greenhouse differs from the sunspace described above in its allowance of room for planting beds and gardening activities. The depth of the greenhouse is based on the necessary width of the planting beds, the corridor, and the storage and work spaces. For example, the planting beds, along the south wall, might be 2 feet in depth, the corridor 2 to 3 feet wide, and the storage and work space another 2 feet, giving the greenhouse a depth of at least 6 to 7 feet. Greenhouses require vents to avoid overheating plants, since temperatures may rise beyond plant tolerance levels. Thermal shutters should be fitted to the south-glazed section, between rafters or structural supports, since you will want to hold heat in the greenhouse as well as in the house. These shutters must be weather-stripped and airtight when they are closed.

Materials in a greenhouse should be moisture-resistant. Cedar framing members and trim serve well, but cedar is becoming rare and is shipped in at high energy cost from Canada and Washington. Pine or fir requires use of a preservative to protect the wood. Check with local paint dealers to determine the most suitable material for this special use, since some preservatives might be toxic to plants.

A solar attic is a sunspace built above the wall line of the building, as a part of the normal roof configuration. This method is very useful where neighboring buildings, hillsides, or trees prevent the use of systems at floor level. Especially in tight urban areas, a solar attic may be the only logical choice. The south slope of the roof is glazed rather than shingled, and the south half of the attic

Below, left: The greenhouse does double duty as a heater and a garden.
Below, right: An earth-bermed greenhouse. Earth piled against the lower walls helps to maintain nighttime temperatures high enough to protect the plants in winter.

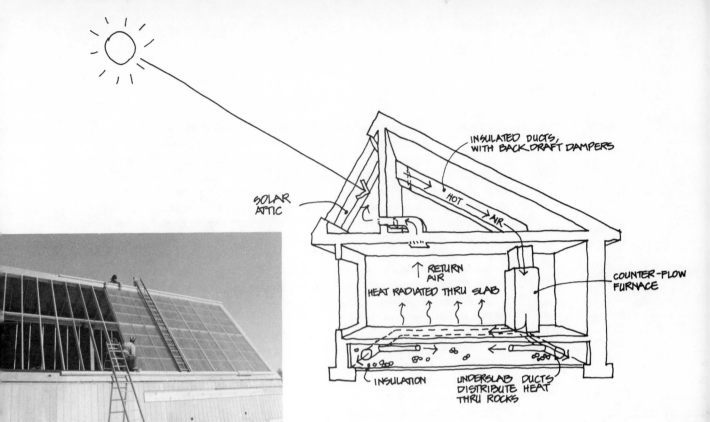

Labels in diagram:
INSULATED DUCTS, WITH BACK DRAFT DAMPERS
SOLAR ATTIC
HOT → AIR
↑ RETURN AIR
HEAT RADIATED THRU SLAB
COUNTER-FLOW FURNACE
INSULATION
UNDERSLAB DUCTS DISTRIBUTE HEAT THRU ROCKS

Above, left: A solar attic at Soldiers Grove, Wisconsin, under construction.
Above, right: How a solar attic collects heat and, in conjunction with rock storage beneath the ground floor, warms the interior space.

is partitioned off from the rest of the attic and drywalled at walls and floors. The walls and floor are insulated and painted black on the interior. Since this space is separated from living areas, no thermal shutters are needed.

A fan is needed to pull heat from the solar attic into the living space. If the attic is hooked into a backup heating system, the same ductwork and fans can serve both systems, carrying heat directly into the living space, or into a remote storage system for return to the living space as needed. Dampers in the ducts close off the attic at night and on cloudy days to prevent cold air from dropping, or "ponding," into the living space.

Materials for sunspaces are those common to other methods. Glazing, for instance, can be regular window glass, low-iron-content glass, or fiberglass—just as for the mass heat wall. Be careful to allow for expansion and contraction of the glazing material. Glazing mullions must be installed in a good workmanlike manner to avoid leakage. Neoprene expansion joints between glazing components may help to prevent leaks.

A sunspace can be designed to double as a screened porch in summertime, with the addition of shutters at the roof to reflect the heat out and with operable vents covering large areas of the south, east, and west walls, to open the space up to cooling breezes. Some of the south-facing glazing panels could also be operable, like a window, or removable, to admit further breezes.

Vents at the roof line in sunspaces, coupled with low vents at the floor line in the south wall, can be used to remove heated air in summer, preventing overheating. These vents can be operated manually or with spring-loaded canisters, such as we have described in the mass-wall discussion. Natural convection will pull heated air out of the space, or you can use fans powered by photovoltaic, or solar-electric, cells. These little fans work when the sun is shining and excessive heat buildup is occurring. (You will have to plan the installation of the fan so it won't operate when you want heat in the house; for instance, a warm 60° day in December could trigger the vent. If this happens, you can override the device manually to keep your solar heat.)

Turbine vents can be installed at the top of the roof to pull heat out of the sunspace in the summer. They have fluted openings that catch the wind. The wind causes them to turn, and this movement pulls air out of the space. Install these vents so that they can be dampered off or entirely closed off in the winter, when you want heat.

Sunspaces are an attractive and effective method of retrofitting an older building. They will simply be an extension of the building on the south, with the appearance of a large window wall or a greenhouse, one or two stories high. A greenhouse seems especially suitable for an addition, with the many ways that it can enhance your life—growing food and flowers, extending your kitchen, giving you a breakfast room, garden room, hot-tub room—so many exciting possibilities. As for decorating or maintaining these sunspaces, we can't think of any problems. Of course, there are a lot of windows to wash, but surely the natural moisture of the rain will help a bit.

The sizing of sunspaces, including all variations, is identical to direct-gain sizing. To achieve 90% to 100% of your annual space-heating requirement in a thermally efficient house, you will need 25% to 100% of the total square footage of your living space in area of glazed sunspace, depending upon where you live. As for storage of heat, the options vary as much as sunspace strategies.

For a greenhouse or sunspace at the first floor, for instance, storage may be in a thermal mass, such as a concrete floor slab in the space, requiring for each square foot of glazing 200 pounds of mass for direct gain and 600 pounds for indirect gain. Insulating shutters will be needed to protect the storage mass from cold outside temperatures at night.

A greenhouse, built on a slab of concrete with concrete walls as thermal mass, shown under construction.

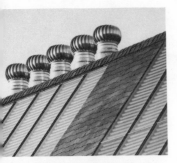

Above: These turbine
vents provide ventilation
by drawing air from
within the house.
Below: This house will
have a rock storage
system underneath the
concrete slab floor. The
rocks should be clean
and well-rounded,
to maximize thermal mass
and allow air circulation.

In a solar attic, a heavy storage mass is not practical, since its weight presents structural problems. The expense of building a structure strong enough to support the thermal mass is not economical for a typical house. Instead, the heat may be held in the attic space and drawn from the space as it is needed to heat the house. This method is simple and direct. It will supply heat whenever the sun's warmth is being collected in the space and will continue to provide heat after sunset, until the heated air has been dissipated.

Another way to store heat from the solar attic, and from any of the other sunspaces, is to connect it by means of conventional heat ducts to a remote storage area of rock and concrete block. The storage area may be in a corner of the basement or in a crawl space. This is how it works: Concrete blocks are laid with their openings lined up in sequence to form air channels at the bottom of the bed. They are covered with wire mesh and then with rock. The rock should be rounded and ideally 1½ to 2 inches in diameter, which allows for the best combination of air circulation and thermal mass. An air space at the top completes the bed. The enclosure is a bin built of conventional insulated stud walls. Heat flows into the rock bed through ducts from the sunspace and is drawn out as needed to heat the house. Heat from a sunspace can also be stored in a bed of rock below a concrete floor, with the heat being conducted directly through the floor to the living spaces or vented through floor registers.

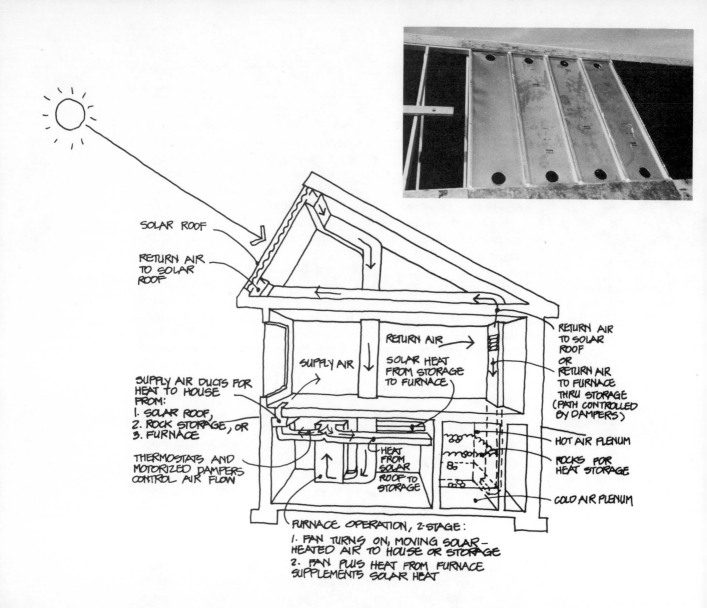

SOLAR ROOF

RETURN AIR TO SOLAR ROOF

SUPPLY AIR

RETURN AIR

SOLAR HEAT FROM STORAGE TO FURNACE

RETURN AIR TO SOLAR ROOF OR RETURN AIR TO FURNACE THRU STORAGE (PATH CONTROLLED BY DAMPERS)

SUPPLY AIR DUCTS FOR HEAT TO HOUSE FROM:
1. SOLAR ROOF,
2. ROCK STORAGE, OR
3. FURNACE

THERMOSTATS AND MOTORIZED DAMPERS CONTROL AIR FLOW

HEAT FROM SOLAR ROOF TO STORAGE

HOT AIR PLENUM

ROCKS FOR HEAT STORAGE

COLD AIR PLENUM

FURNACE OPERATION, 2-STAGE:
1. FAN TURNS ON, MOVING SOLAR-HEATED AIR TO HOUSE OR STORAGE
2. FAN PULLS HEAT FROM FURNACE SUPPLEMENTS SOLAR HEAT

Occasionally we find a building site that is on a south-facing slope, and in some of these cases we have built a sunspace below the floor line of the house. In this situation, heated air from the sunspace will rise naturally, and we channel it through the concrete blocks and storage bed without the assist of a fan. We did this at the Pleasant Valley Outdoor Recreation Center in Woodstock, Illinois, using such a sunspace in combination with an earth-integrated structure.

The amount of rock needed for remote storage is difficult to measure in terms of weight. Instead, we say that for every square foot of collector, you will need half a cubic foot of storage rock.

Solar Roof. Like the solar attic, the solar roof is built as an integral part of a south-facing roof. The roof angle is designed to be as close as possible to perpendicular to the sun's rays during the critical winter heating months of December, January, and February. This angle should be equal to the latitude where the building will be constructed plus 10° to 15°. Thus, at 40° latitude, the angle would be 50° to 55°.

Built on site with simple materials, the solar roof is integrated into the roof joists of the house. While the solar attic is a sunspace at roof level, the solar roof differs in its construction and its operation. It is built in the following manner: Metal pans, 1½ inches deep, are placed between the rafters. On top of them is placed a layer of corrugated metal roofing material painted black. Two layers of glazing go over the corrugated metal. When the sun strikes the blackened metal, it is heated and in turn heats the air surrounding it. Removed by a fan from the space between the corrugated metal and the pans, this air is used to heat the house.

The interesting thing about this system is that each layer of material substitutes for a layer of the customary materials used in construction of a roof. As a result, we find that buildings using this system can be put up for the same square-foot cost as any other building.

Opposite page: Metal pans for a solar roof, with air-intake and exit holes at bottom and top, shown in place between the roof rafters (top). How a solar roof works, in conjunction with rock storage in the basement of the house (center). Black-painted corrugated barn roofing and fiberglass panels are nailed down over the metal pans (bottom) to complete a solar roof. This page: The Selby House, Osseo, Wisconsin, displays its solar roof.

Solar roof systems require fans to move the air around, which uses additional energy. The other systems described are frequently coupled with fans, too, and the energy cost is about the same in all cases. Heat is stored in a bed of rock, remote from the system, either in a basement, crawl space, or under the concrete slab on the first floor of the house. This bed is identical to the one described in sunspace storage.

Sizing for these site-built collection systems is similar to that for other methods. For 90% to 100% of the total heating requirement, you need about 25% to 100% of the floor area in collection surface. For every square foot of collector, you need half a cubic foot of storage rock.

In all of the systems, you will note that the solar-heat components of the building, like all other building components, must be properly built, with tight and carefully fitted workmanship. Dampers must be airtight. All ductwork must be well-insulated and airtight, as must blower cabinets and furnaces. In fact, "airtight" is the word for everything in the job. Rock storage bins, collectors, glazing, air locks—all connections should be tight so that cold and heated air do not mingle. Insulation should be installed carefully in walls, ceilings, shutters, and all other places where it is used to avoid air leaks.

Any solar-heating system that you might choose will be only as good as the workmanship that is brought to the construction process. We have found that, when problems develop, they have simple causes. An air leak, a piece of loose insulation, a damper that does not close tightly—these are the difficulties. Careful

The Sobol House, McHenry County, Illinois, combines direct-gain windows with a solar roof and a greenhouse.

The Sobol House

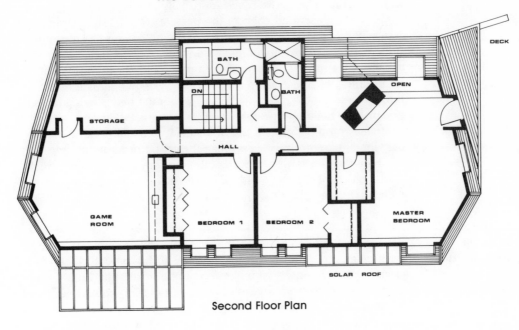

Second Floor Plan

DECK

BATH

DN

STORAGE

OPEN

BATH

HALL

GAME ROOM

BEDROOM 1

BEDROOM 2

MASTER BEDROOM

SOLAR ROOF

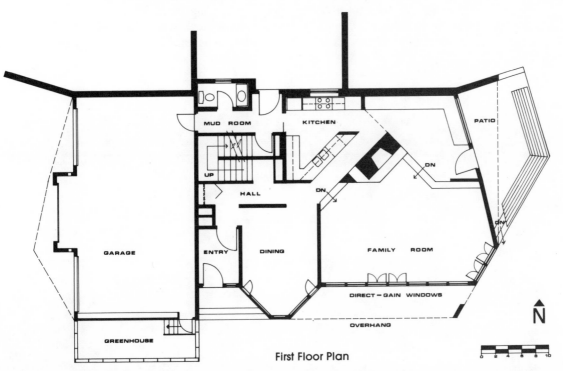

First Floor Plan

MUD ROOM

KITCHEN

PATIO

UP

DN

HALL

DN

DN

GARAGE

ENTRY

DINING

FAMILY ROOM

DIRECT-GAIN WINDOWS

OVERHANG

GREENHOUSE

N

0 2 4 6 8 10

attention to details during construction will help to prevent poor workmanship. This is discussed more fully in Chapter 6.

We have described each of the passive solar-heating systems as a separate entity, but in practice we think of them as a group of strategies for heating buildings. As a result, we frequently—in fact, usually—use more than one system in a building. A combination, for instance, of direct gain and solar roof will give instant heat through south-facing glass while a rock storage bed is being heated to carry the building through the night. A direct gain and mass heat wall combination can provide similar versatility. For gardeners, a greenhouse is a must; it may be used alone or combined with a mass wall or another system. Each project, each site, each client is a new and different challenge, and we orchestrate the solar design accordingly.

Glazing Requirements

Throughout our discussion of passive solar-heating systems, we have given you only rough guidelines as to how much solar glazing you can expect to build into a new or existing home. The guidelines varied from an amount of solar glazing equal to 25% of the floor area, to an amount equivalent to its entire size. In the next chapter, we will go into detail as to how to make your own calculations, for your individual location and kind and size of house. In the meantime, however, we have worked out "rule of thumb" guidelines for thermally efficient homes in 31 cities

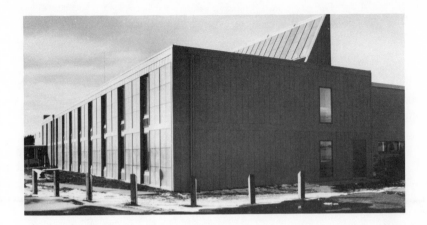

Opposite page and this page: The Alan House Motel combines direct-gain windows, mass heat walls, and solar roofs to satisfy up to 100% of its heating requirements.

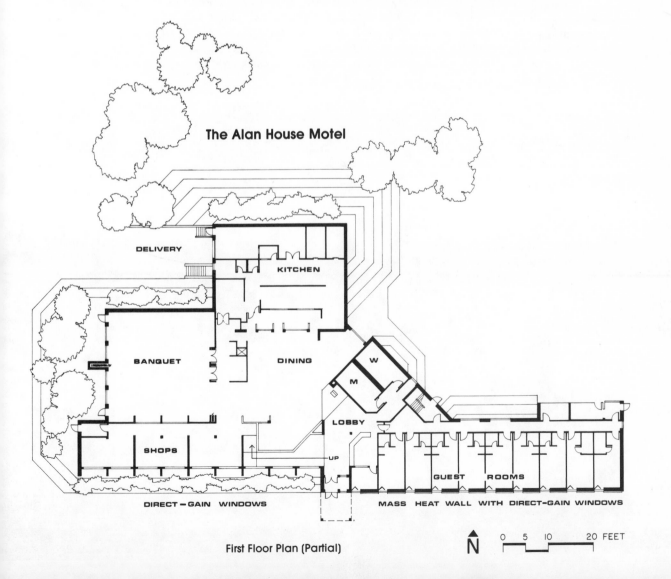

The Alan House Motel

DELIVERY

KITCHEN

BANQUET

DINING

W

M

LOBBY

SHOPS

UP

GUEST ROOMS

DIRECT-GAIN WINDOWS

MASS HEAT WALL WITH DIRECT-GAIN WINDOWS

First Floor Plan (Partial)

N   0 5 10   20 FEET

around the United States. If you follow our instructions on thermal efficiency and live in or near one of these cities, the amount of glazing the table on page 93 calls for should work for you.

Natural Cooling Strategies

If your house gets too hot for comfort in summer, can you cool it naturally? According to the analyses of macroclimate in 31 cities presented in Chapter 1, in most cases when high temperatures and high humidity combine to provide weather conditions above the comfort range, cooling can be accomplished by use of the wind and shading. We have discussed briefly the ways in which your heating system can help to cool the house; now let's look at cooling in more detail.

All of the strategies that we have described to make a thermally efficient building will work for you to keep your house cool in summer, just as they work to keep it warm in winter. One of our own buildings, built on an old barn foundation, has earth-integrated walls that are 5 feet thick at the base and 2½ feet thick at the top, with 12 inches of insulation in the roof and 6 inches in the walls. In summer

The Wright House, near Osseo, Wisconsin (shown under construction), uses an old barn foundation as the walls of its first floor. The thick stone walls, earth-integrated, help keep the house cool in summer.

Glazing Requirements for a Thermally Efficient House

City	Amount of Solar Glazing in Proportion to Floor Area	Annual Total	Solar Contribution to Heating Needs — Monthly Percentage											
			J	F	M	A	M	J	J	A	S	O	N	D
Asheville, NC	25%	99%	99	100	100	100	100	100	*	*	100	100	100	99
Boise, ID	25%	95%	89	100	100	100	100	100	*	100	100	100	100	92
Boston, MA	25%	93%	88	95	97	100	100	100	*	100	100	100	100	90
Burlington, VT	100%	92%	87	94	97	98	100	100	100	100	100	100	98	83
Chicago, IL	25%	93%	90	96	98	100	100	100	*	*	100	100	100	89
Denver, CO	25%	99%	98	99	99	100	100	*	100	100	100	100	100	99
Des Moines, IA	25%	91%	85	95	96	100	100	100	*	*	100	100	100	87
Detroit, MI	50%	94%	89	97	98	100	100	100	*	100	100	100	100	89
Dodge City, KA	25%	99%	98	100	100	100	100	100	*	*	100	100	100	98
Duluth, MN	100%	90%	83	94	96	96	97	100	100	100	100	100	95	79
Fargo, ND	100%	91%	83	96	97	99	100	100	100	100	100	100	97	83
Green Bay, WI	50%	91%	84	94	96	99	100	100	100	100	100	100	98	83
Harrisburg, PA	25%	94%	91	97	99	100	100	*	*	*	100	100	100	91
Indianapolis, IN	25%	91%	86	95	98	100	100	100	*	100	100	100	100	86
International Falls, MN	100%	86%	76	93	96	97	99	100	100	100	100	100	91	73
Louisville, KY	25%	97%	94	99	100	100	100	100	*	*	100	100	100	95
Madison, WI	33%	90%	84	94	95	100	100	100	100	100	100	100	98	81
Minneapolis/St. Paul, MN	50%	90%	83	94	96	100	100	100	100	100	100	100	98	81
New York, NY	25%	94%	91	96	98	100	100	*	*	*	100	100	100	92
Omaha, NE	25%	91%	83	95	96	100	100	100	*	*	100	100	99	87
Portland, OR	25%	96%	93	100	100	100	100	100	100	100	100	100	100	95
Rapid City, SD	25%	92%	85	95	95	100	100	100	100	100	100	100	99	88
Reno, NV	25%	99%	98	100	100	100	100	100	100	100	100	100	100	98
Richmond, VA	25%	99%	99	100	100	100	100	100	*	*	100	100	100	99
St. Louis, MO	25%	97%	95	99	100	100	100	100	*	*	100	100	100	96
Salt Lake City, UT	25%	96%	93	100	100	100	100	100	*	100	100	100	100	94
Seattle, WA	25%	94%	89	100	100	100	100	100	100	100	100	100	100	90
Spokane, WA	33%	90%	81	99	100	100	100	100	100	100	100	100	100	80
Springfield, MO	25%	98%	97	100	100	100	100	100	*	100	100	100	100	98
Syracuse, NY	50%	90%	86	93	96	100	100	100	100	100	100	100	100	83
Youngstown, OH	50%	92%	87	94	96	100	100	100	100	100	100	100	100	86

*Heating not required.

The vents below
south-facing windows
draw cool air into
the house.

the building stays 15° below outside temperatures during even the hottest weather, because earth integration moderates the temperature extreme and insulation reduces heat gain, just as they reduce heat loss in winter. Much of the cooling requirement can be satisfied, then, by building a thermally efficient house, especially an earth-integrated one.

You can achieve further cooling through proper ventilation. Operating windows can be opened to bring in fresh air, with high vents or windows removing heated air from the house. Building codes usually require that there be ventilation equivalent to 5% of the floor area of the house. This is generally sufficient to cool the building comfortably, if the windows and vents are correctly arranged.

The effectiveness of ventilation for cooling depends on the placement of the openings. The natural forces available for cooling with ventilation are wind speed and temperature differential. Windows or vents that admit air to the house should face the prevailing winds, with some facing in other directions from which summer breezes may flow. Summer winds in the Midwest tend to be out of the south and southwest, but local variances should be checked and accommodated. Vents that remove heated air should be high and on the opposite side of the house from the inlet windows.

For example, a monitor, skylight, or clerestory might be 10 feet higher than the windows admitting air. This creates a stack effect, pulling cool air into the house and sending heated air out through high vents at the same time. At night, the stack effect works especially well, as there is usually the greatest difference between indoor and outdoor temperature at this time. The more you can cool down your house during the night, the cooler it will remain the following day. It is important that you avoid windows on the west, especially in bedrooms, for when the sun is setting in the northwest in the summer, it will overheat these rooms.

You will remember from the climate analyses that in all 31 cities shading is required in summer months—and sometimes in spring and fall—to keep direct sunlight out of the living space. The best way to block the sun is with wide roof overhangs. Using the angle of the midday sun at the summer solstice, you can determine how wide the overhang needs to be to keep the high summer sun from striking your south-facing wall. Proper placement of deciduous shade trees will also help. Another method of blocking the sun is with thermal shutters. Painted white on the window side, they will reflect the sun away from the windows.

If none of these strategies is enough to keep you comfortable in summer, you may have to resort to more complex methods, such as ground-water circulation and air-earth tubes. A tavern in Soldiers Grove, Wisconsin, for example, has used ground water to cool the building for 30 years, by circulating this constant low-temperature water through a heat exchanger in the ventilation ductwork. The air-earth tube has been employed in the Middle East for centuries and is being used successfully in Iowa for cooling hog barns. Air is drawn into a metal culvert tube buried 7 to 10 feet below grade and circulated through the building by a fan. The condensation that occurs in the tube must be taken off by a sump or a drywell. The tube must be screened to keep out rodents, insects, and pets, and it must be planned so it can be closed down tightly in winter when you do not want cold air entering your house. Studies now being conducted, based on computer models, may give us more scientific background than the Middle Easterners had, but we already know that air-earth tubes work.

Use of these strategies, and proper management of them, should provide a cool and comfortable summer. Remember, though, that

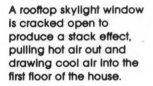

A rooftop skylight window is cracked open to produce a stack effect, pulling hot air out and drawing cool air into the first floor of the house.

there will be some times when the temperature extreme and the relative humidity extreme coincide and nothing you can do will thoroughly cool the house. These are brief incidents, of no real consequence when viewed within the scope of the entire summer.

Hot-Water Heating   Taking care of space-heating needs is all very well, you say, but my family uses an incredible amount of hot water. Showers average 15 minutes each, and all family members take at least one a day. Help!

We try to solve this problem by integrating hot-water heating with space heating. In a sunspace, for instance, it could work like this: Stripped of their jacket and insulation on the side to be exposed to the sun, two hot-water storage tanks are mounted high in the sunspace, between the glazing and the inner wall of the house. These tanks, painted black, are connected in series to the plumbing system. Cold water from the water service enters the tanks and is heated by the sun. At the time of use, water in the second tank is hotter than the water in the first one. Cold water replaces water sent from the first tank into the second one, and it in turn is heated. You can

Solar hot-water tanks shown in place on the roof of the Anderson House.

protect the tanks from freezing at night and on cold, cloudy days with insulating shutters similar to the ones you use for windows. They must be completely airtight.

These tanks can also be mounted underneath and in front of the waist-high planting beds in a greenhouse, doing double duty by storing heat for cold nights. Be sure that they are not shaded by the planting beds.

If your heating choice is a mass heat wall or direct gain, you can build a small sunspace just for heating water, locating it high on the south wall or integrating it into the south roof of the house. Place tanks in this water-heating sunspace as you would in the larger sunspace. Or, as another alternative, you can place the tanks high in the house, especially in a two-story house—perhaps at the top of a stairwell—so you can use rising heated air within the house to heat your water.

Two 35-gallon tanks will satisfy the needs of an average family. Note that these are *water-storage* tanks. This is very important. These tanks are built just like water-heater tanks, with a glass lining to reduce corrosion. They should be installed with gate valves, a kind of valve that makes it possible to bypass the tanks when you want to and when you need to make repairs. A hose bibb should be included at the service entrance to the tanks, so they can be drained. All of the normal plumbing codes must be satisfied in this work.

You may wish to install a conventional water heater as well, for those times when the sun will not give you the desired temperatures. The water is preheated in the sun heater and then passes into the water heater, which completes the job.

With the solar roof and solar attic, you can heat water in the duct that brings hot air from the collector to the storage bed. Cold water runs through a heat exchanger, similar to a car radiator, circulated by a small pump. A differential thermostat controls this heating system. When sufficient heat is not available, the pump will not operate and the backup hot-water heater will do the heating. You must install an airtight damper between the heat exchanger and the collector, to prevent freezing during cold nights and cloudy days.

Illuminating a Solar House

How can you illuminate a passive solar house without wasting energy? As much as possible, use natural lighting—that is, daylight. Even on cloudy days, the daylight will be bright enough in a properly designed house so that you won't have to use artificial lighting.

Solar design, with its south-facing windows, creates the best

Solar houses are bright and cheerful during the daytime, reducing the need for artificial illumination.

orientation for daylighting the house. To make optimum use of daylight in a room, the distance from the windows to the opposite wall should be no more than 20 feet. If the room is deeper than 20 feet, you can use skylights or clerestory windows to provide a second source of natural light. Windows placed so that they intersect interior walls will permit light to be reflected off the wall, thus increasing the brightness of the room. In fact, daylighting is so effective that summer shading is necessary to reduce the intensity of the direct sun, and work areas may have to be oriented away from the light source to avoid distracting glare. Thermal shutters can be used selectively to adjust daylight brightness levels.

Providing sufficient daylighting is not difficult. Most building

Direct-gain windows at the Theios House bring sunlight into the deepest part of the room in winter.

codes require a window area equal to 10% of the floor area. In a 20-by-20-foot room this amount, placed on the south side of the house, provides the following levels of illumination at noon in December: 42 footcandles at 5 feet from the window, 9 footcandles in the middle of the room, and 9 footcandles 5 feet from the back wall. For comparison, 3 footcandles are required to read a book.

The decorating of the house plays an important role in daylighting. Floors and furnishings should be light and bright. Ceilings should be 80% reflective and walls should be 50% to 60% reflective. We accomplish this by painting drywall white or by using light stains or preservative on light-colored woods.

A house planned for daylighting does not need artificial lights until dusk, when the world outside begins to grow dark. Our approach, as with heating, is to use the least amount of energy for lighting the house. This can be done in two ways: through the use of low-energy fixtures and through the use of "task lights," or lights placed at the proper locations. Since fluorescent bulbs provide nearly five times more light per watt than incandescent bulbs, we recommend the lowest-wattage fluorescent bulbs possible, in the least numbers that will provide adequate light for the house. Integrated into valances or cabinets with the surrounding enclosure painted white for maximum reflectance, these lights are unobtrusive and effective.

Since the intensity of lighting varies inversely with the square root of the distance traveled, a brighter light is required to read a book if the light source is placed at the ceiling rather than directly over the desk where the reader sits. The least amount of energy will be

Right: Mirrors reflect natural daylight into the room, while a wooden valance covers an indirect fluorescent light fixture. Note the space-saving fold-down table. Below: Fluorescent task lighting.

consumed if lighting for specific tasks is located where the task is done, using low-wattage fluorescent fixtures with glare-reducing lenses. These may be built into cabinets or under shelves, or they may be wall-mounted directly above a desk, kitchen counter, or workshop bench.

Small, well-placed windows amply illuminate the second floor of this solar house. The ceiling fan at the top of the stairs recirculates warm air in winter and provides movement of air in summer.

To avoid eye fatigue, you also need some general lighting to illuminate the surrounding room. During the day this can be natural light only. At night you can use a single-bulb fluorescent fixture, wall-mounted near the ceiling with a wooden valance, to wash the ceiling and walls with continuous background light.

This reduction of demand for artificial lighting is just another example of the ways in which your solar house can save energy while still serving you with the comfort to which you are accustomed.

Calculations

4

In the preceding chapters we have described methods used to achieve thermal efficiency and have given some guidelines for the amount of solar glazing necessary to maximize the solar contribution in such a thermally efficient house.

In this chapter we will quantify our recommendations. This involves a lot of arithmetic. Why is it necessary? Why not just follow recommendations and install solar glazing and hope for the best? You can. But in many cases your house will be different from those we've described. Maybe your existing house cannot be insulated heavily enough. How much solar glazing should you use then? Maybe you cannot fit enough glazing on your house. How much of your heat will be provided by the sun? Or maybe you need to be convinced with figures that the recommendations we have made are necessary for your new house.

Calculating Thermal Efficiency

Thermal efficiency is a measure of the amount of energy required to maintain comfort in a building. The methods we have described can result in a building using only 4 British thermal units (Btu's) or less a day for each square foot and for each Degree Day, or 4 Btu/DD/sq. ft. In comparison, a poorly insulated house may use as much as 20 Btu/DD/sq. ft. Once you have calculated the thermal efficiency, you can estimate the amount of energy required to heat the house for a year. Then you can estimate how much of that heat can be provided by the sun.

To measure the thermal efficiency of a house, you must first

102

An airtight wood stove is an excellent backup heater for a solar house.

determine its heat loss. Heat-loss calculations serve as a guideline to indicate the level of compliance with thermal requirements. You begin by finding out how many Btu's escape from your house through each exterior element of the building—windows, walls, roof, doors, and floors.

Heat loss through this envelope of the house is known as conduction heat loss—that is, heat conducted through the building materials to the cold outside. To prevent conduction heat loss, we use layers of materials that will impede the flow of heat. This resistance is measured in R Values, which indicate the resistance to heat loss. You will see an R Value shown on the outer package of insulation in the lumberyards. R Values for other materials are listed in sources such as the American Society of Heating, Refrigerating, and Air-Conditioning Engineers (ASHRAE) *Handbook of Fundamentals,* which you can find in most libraries. Table 1 shows what each layer of material in a typical wall of a thermally efficient house offers in resisting the loss of heat.

Table 1: R Values
for Typical Components of a Thermally Efficient Wall

Wall Component	R Value
Air film on exterior surface (15 mph average)	0.17
Siding	0.81
Paper	0.06
Sheathing	1.32
12-in. batt insulation	38.00
4-mil vapor barrier	negligible
⅝-in. drywall	0.46
Still air on interior surface	0.68
TOTAL	41.50

R Values represent *resistance* to heat loss; to calculate the actual rate of heat loss through a particular part of the house envelope, R Values are converted to U Values, which express heat loss in terms of Btu's per square foot per degree Fahrenheit per hour. To find the U Value, you divide 1 by the R Value. For the wall described in Table 1, the U Value equals 1 divided by 41.50, or .024. This means that when outdoor temperature is lower than indoor temperature, each square foot of the wall will lose .024

Btu's per hour for *every* degree of difference.

You can find the U Value for some building components in the ASHRAE *Handbook,* using the formula above. The lower the U Value, the higher is the thermal efficiency for the component. Table 2 shows how different thicknesses of insulation, as well as various levels of thermal insulation for windows and doors, will dramatically affect U Values.

Table 2: U Values for Components of the House Envelope

Building Component		U Value
WALLS: Wood frame with wood siding or 4-in. face brick	with no insulation	.220
	with 3½ in. insulation	.069
	with 6 in. insulation	.044
	with 12 in. insulation	.024
CEILING: Wood frame with drywall or plaster	with no insulation	.280
	with 6 in. insulation	.046
	with 12 in. insulation	.025
	with 24 in. insulation	.013
FLOOR: Wood frame with oak flooring	with no insulation	.450
	with 3½ in. insulation	.075
	with 9 in. insulation	.032
WINDOWS: Wood frame	single-glazed	1.130
	double-glazed	.580
	triple-glazed	.360
	triple-glazed with shutters	.100
DOORS: Solid wood core	without storm door	.490
	with storm door	.270

Hourly heat loss through each component of the house envelope is found by multiplying the U Value by the area of the surface (wall, ceiling, floor, window, or door), and then multiplying that figure by the ΔT, pronounced "delta tee." The result is expressed in Btuh, or Btu's per hour. The ΔT is the difference between the indoor thermostat setting (normally 65°) and the design temperature. This design temperature (for heating) is the minimum above which the outdoor temperature will remain 97.5%

of the time; it can be found for your location in the ASHRAE *Handbook.* Thus, ΔT equals the indoor thermostat setting minus the design temperature.

Table 3 represents a sample problem, calculating heat loss for a 1,000-square-foot house (20 feet by 50 feet), each component of which has the maximum amount of thermal efficiency shown in Table 2. The areas for the various parts of the house are:

> Ceiling—1,000 sq. ft.
> Floor—1,000 sq. ft.
> Walls—978 sq. ft. (8 ft. height × 140 linear ft. of
>   walls minus door and window areas)
> Windows—100 sq. ft. (10% of floor area)
> Doors—42 sq. ft. (two entries, each 3 ft. by 7 ft.)

Let's assume our house is in Chicago, where the design temperature is 0°; therefore, the ΔT is 65° (the indoor thermostat setting minus the outdoor design temperature, or 65°−0°).

If you've come this far with us, take heart! The worst is over.

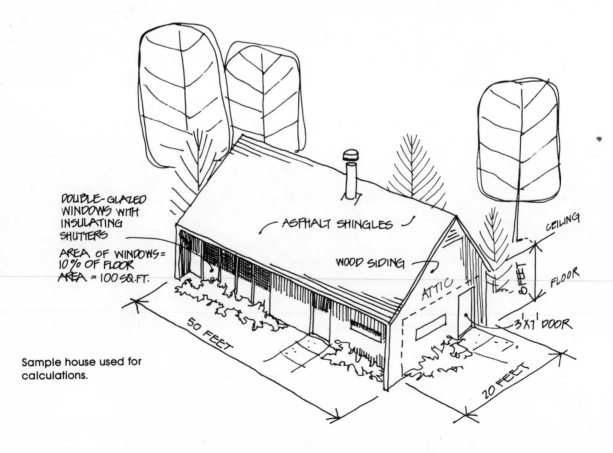

Sample house used for calculations.

Table 3: Heat Loss
Through House Envelope (Conduction Heat Loss)

Part of House	U Value (Btu/sq. ft./°F/hr.) ×	Area (sq. ft.) ×	ΔT (°F) =	Heat Loss (Btuh)
Walls	.024	× 978	×65 =	1,526
Ceiling	.013	× 1,000	×65 =	845
Floor	.032	× 1,000	×30* =	960
Windows with thermal shutters	.100	× 100	×65 =	650
Doors with storm doors	.270	× 42	×65 =	737

TOTAL 4,718

*Here the ΔT is different because we assume that the crawl space is warmer than the outside temperature, since it is protected by the earth. We assume ΔT = 65°F − 35°F, or 30°F.

In addition to heat lost by conduction through the building's envelope, heat is lost when cold outside air comes in through cracks and must be heated. The entry of cold air, called infiltration heat loss, is especially common around windows and doors. That is why careful construction, weather stripping, and caulking are so important—to minimize these cold drafts.

There are several ways to calculate infiltration heat loss. First, you must estimate the amount of air entering the building. Some air change is necessary for comfort and health, to prevent stuffiness and buildup of odor and moisture. In a tight, thermally efficient house, the amount of air entering through cracks on the windward side, and exiting through cracks on the leeward side, is roughly equal to three-quarters of the volume of the heated space per hour. (In our practice, we assume that one-half of the air in the house is replaced each hour, but we will use the more conservative figure of three-quarters of the volume for these calculations.) This means that three-quarters of the air in the house is replaced each hour. The volume (V) is equal to the ceiling height (H) times the length (L) and width (W) of the floor, or: $V = H \times L \times W$.

To calculate infiltration heat loss, the number ¾ V is multiplied by

Backup Heating Systems: When the Sun Doesn't Shine

Even the most thermally efficient of passive-solar-heated houses needs a backup heating system. The sun may contribute well over 90% of the annual heating requirement, but this is an average figure, and there will still be days when the sun does not shine and outdoor temperatures are extremely low. The peak heat load, which we calculate in this chapter, is an indication of how many Btu's per hour (Btuh) may be needed to keep you comfortable during these crucial times of the year. If the sun cannot do the job, then it is up to your backup heating system to do it.

Passive solar houses may use one of a variety of backup systems. Airtight wood-burning stoves, of various designs and sizes, can be excellent auxiliary heating devices. Wood-burning furnaces are also available for those who prefer a central heating system. Wood burners must be safely installed and carefully used; we suggest that you follow the recommendations of the National Fire Protection Association.

You may prefer an automatic backup system — a furnace which uses gas, oil, or electricity as its fuel — if you expect to be away from home a good deal or if you don't choose to tend a wood-burning heater. You may also want to investigate multi-fuel furnaces, which use wood as well as one or more other fuels. When the wood fire dies down, an alternate fuel will provide heat.

The high thermal efficiency of your passive solar house will result in a small heating requirement for your backup furnace. A unit sized to meet this low Btuh demand may be difficult to find.

For a solar-heating method that uses a fan to draw warm air into the house, a forced-air furnace could provide the fan for the solar heat, as well as serving as the backup heating system.

ΔT times .018, which is a kind of U Value for the cold air infiltrating a warm house. The formula is:

$$\text{Infiltration heat loss} = .018 \times \tfrac{3}{4}\, V \times \Delta T$$

For the sample house described for Table 3 (ΔT, remember, was 65):

$$V = 8 \text{ ft.} \times 50 \text{ ft.} \times 20 \text{ ft.} = 8{,}000 \text{ cu. ft.}$$
$$\tfrac{3}{4}\, V = 6{,}000 \text{ cu. ft.}$$

$$\text{Infiltration heat loss} = .018 \times 6{,}000 \times 65 = 7{,}020 \text{ Btuh}$$

For reference, when you are dealing with a house that does not have excellent thermal efficiency, you will have to estimate a higher infiltration of air. This will require more extensive study of the house in question in order to determine the actual infiltration rate. Your architect, a mechanical engineer, or the ASHRAE *Handbook* can help you make this determination. Example 3 at the end of this chapter addresses this problem.

There is another thing to note about a thermally efficient house versus a "leaky" house. When a house is designed for a minimum amount of infiltration, air-intake ducts should be provided for combustion appliances: gas, oil, or wood-burning stoves and furnaces. The leaky house does not need these special fittings—the air for the stove or furnace is coming in through the cracks! Don't worry about not getting enough air in your thermally efficient house, however—fresh air can be supplied through operable windows and vents at any time.

Now, to find the total hourly heat loss for the sample house, add the conduction and infiltration heat losses together:

4,718 Btuh	conduction heat loss
+7,020 Btuh	infiltration heat loss
11,738 Btuh	peak heat load

The total, called the peak heat load, represents the maximum rate of heat loss on a cold night, when the outdoor temperature is the same as the design temperature. (Remember that outdoor temperature will be higher than design temperature 97.5% of the time.) The peak heat load tells you what the output size of your backup heating system should be. That is, your backup heating system should be capable of supplying enough heat each hour to make up the loss of that much heat.

In order to calculate heat loss for a year (which, among other things, will let you estimate your conventional heating bill), you must first determine the loss for each Degree Day, or Btu/DD. This may be calculated using the formula:

$$\text{Btu/DD} = \frac{\text{Btuh (peak heat load)} \times 24 \text{ (hrs./day)}}{\Delta T}$$

For the sample house:

$$\text{Btu/DD} = \frac{11{,}738 \times 24}{65} = 4{,}334$$

Multiplying the Btu/DD figure by the yearly number of Degree Days for the location will give you the total annual heat loss. Thus, for the sample house in the Chicago area:

$$4{,}334 \text{ Btu/DD} \times 6{,}127 \text{ DD/yr.} = 26{,}554{,}418 \text{ Btu's}$$

total
annual
heat loss

You could stop here, but you'd be shortchanging yourself if you did. You see, you can also calculate internal heat gain—all of the heat that is generated inside the building. It's heat, it's energy. Why not count it? To give you an idea of how much heat is generated inside the house, consider this: You yourself give off approximately 300 Btuh when you are lying down, 400 Btuh when you are just sitting around, and many more when you are working. For a family of four that could amount to an average of 1,600 Btuh of internal gain when everyone is at home.

Other contributors to internal gain include your electrical appliances. Table 4 shows typical amounts of heat generated per year by common household equipment. The yearly kilowatt-hour (kwh) figures represent an assumed annual rate of use for each appliance for a family of four.

Looking at any of the appliances listed, you can see the impact this internal gain can have in heating your house. This does not mean, however, that we recommend using lots of appliances and expending large amounts of energy. On the contrary, we think the use of appliances should be kept to a minimum. The first thing to do is to throw away unnecessary gadgets, unless they are hand-powered, and retain only those appliances that are really needed.

An appliance that most of you will want to keep is the water heater, so let's take a look at its potential contribution to home heating. Annually, an average-sized water heater will give off over 16,000,000 Btu's. If it is placed in a living space such as a kitchen

Table 4: Internal Gain from Household Appliances

	Rate of Use (kwh/yr.)	Heat Gain (Btu/yr.)
Refrigerator/freezer	1,137	3,880,581
Electric range/oven	1,175	4,010,275
Toaster	39	133,107
Microwave oven	1,190	4,061,470
Clock	24	81,912
Coffee maker	36	122,868
Hot-water heater	4,800	16,382,400
Color TV (solid state, switched circuit)	440	1,501,720
Stereo/radio	109	372,017
Table radio	60	204,780
Washing machine	1,103	3,764,539
Vacuum cleaner	46	156,998
Iron	156	532,428
Sewing machine	12	40,956
Lighting	600	2,047,800
Task lighting	200	682,600
Convenience outlets	400	1,365,200
TOTAL	6,727	TOTAL 39,341,651

or bathroom, this heat will help warm the house. You can calculate, month-by-month, the internal gain from the water heater during the heating season. In Chicago, where there is a ten-month heating season, you would gain 10,230,167 Btu's of heat per year for our sample 1,000-square-foot, thermally efficient house. Remember: The water heater may give off more heat than that, but sometimes it is unneeded. In months with a low requirement for heat, such as May, June, September, and October, the water heater could supply all of the needed heat, or even more than you would need. In months with a higher heating need, a smaller part of the total heat would be provided.

Obviously the water heater will not be the only source of heat. Heat produced by the occupants, by cooking, and by lighting and appliances will also be present. You can estimate that the total internal gain for your house will be approximately 20,000 Btu's per day per person. For our sample house, let us assume that four people are living in it. Thus four times 20,000 Btu's times 250 days in the heating season gives 20,000,000 Btu's per year of usable heat. We use the length of the heating season because heat is not required—and must be vented—during the remainder of the year. This internal gain figure (20,000,000 Btu's per year) is used to calculate the net heat loss of your home. Net heat loss is total annual heat loss minus internal gain:

$$
\begin{array}{r}
26{,}554{,}418 \text{ Btu/yr. total heat loss} \\
- \underline{20{,}000{,}000 \text{ Btu/yr. internal gain}} \\
6{,}554{,}418 \text{ Btu/yr. net heat loss}
\end{array}
$$

To arrive at a figure for Btu's per Degree Day per square foot, divide the net heat loss by the number of Degree Days per year times the square footage of the house. This will give you the net thermal efficiency of the house:

$$
\text{Net thermal efficiency} = \frac{\text{net heat loss (Btu/yr.)}}{\text{DD/yr.} \times \text{sq. ft.}}
$$

$$
\text{Net thermal efficiency} = \frac{6{,}554{,}418 \text{ Btu}}{6{,}127 \text{ DD} \times 1{,}000 \text{ sq. ft.}}
$$

$$
= 1.07 \text{ Btu/DD/sq. ft.}
$$

How does this net thermal efficiency compare with that of the house next door? Or with that of other houses being built today? To compare, the gross figure for an average house is around 20 Btu/DD/sq. ft. In fact, an all-glass house may be closer to 30 Btu/DD/sq. ft. When we design a house, our goal is a gross thermal efficiency of 2.5 to 4 Btu/DD/sq. ft.

If your calculations tell you that the net thermal efficiency of your house is not good enough, go back to your plan and look for ways that you can reduce the heat loss. Remember: The more you can lower your heat loss, the more solar contribution you will gain—and the less amount of solar glazing you will need.

Sizing Your System

Now it is time to build in the calculations you have been waiting for: how much solar glazing you will need to achieve 90% to 100% of your space-heating requirement. Or, conversely, if you are retrofitting or have limited space for solar glazing, how much contribution you will get from a fixed amount of glazing.

To find the required square footage of glazing for your home, you use the following method. Again, let's use the sample house as an example, to see how the formula works.

For direct gain, mass heat wall, and sunspaces, you must first recalculate the heat loss of the sample house, this time including solar glazing in the plan. (This recalculation is not made for a solar roof or solar attic, since they are isolated spaces, not a part of the heated area of the house.) We will assume that 250 square feet of glazing will be required, as we know from experience that in many places a thermally efficient house will require glazing equal to one-fourth of the floor area, and our sample house has 1,000 square feet of floor space. (Remember that in calculating square footage of glazing for any system, it is important to use net figures. The dimension of any material, such as glazing channels, that interrupts the glazing, must be deducted from the gross area of glazing.) Since the 250 square feet of glazing will be an area of heat gain, rather than heat loss, we deduct 250 square feet from the south face of the wall. Previously, this south wall was calculated to have 100 square feet of windows and 150 square feet of wall; we are now substituting solar glazing for the entire wall area. Therefore, we subtract the following amounts:

$$
\begin{array}{llllll}
& \textbf{U} & \times \textbf{Area} & \times \, \Delta\textbf{T} & = \textbf{Heat loss} \\
\text{Windows} & .100 & \times \; 100 & \times \; 65 & = 650 \text{ Btuh} \\
\text{Walls} & .024 & \times \; 150 & \times \; 65 & = \underline{234 \text{ Btuh}} \\
& & & & 884 \text{ Btuh}
\end{array}
$$

Deducting this from the heat loss we calculated before (11,738 Btuh), gives us an adjusted heat loss total of 10,854 Btuh. This peak heat load is then converted to Btu/DD by multiplying by 24 (hours/day) and dividing by 65 (ΔT):

$$
\frac{10,854 \times 24}{65} = 4,007 \text{ Btu/DD}
$$

Next we deduct internal gain. As we saw earlier, internal gain for this house can be estimated at 20,000,000 Btu's per year. This figure can be converted for use here by dividing by yearly Degree Days (6,127 in Chicago), giving 3,264 Btu/DD.

$$
\begin{array}{ll}
& 4,007 \text{ Btu/DD heat loss} \\
- & \underline{3,264 \text{ Btu/DD internal gain}} \\
& 743 \text{ Btu/DD heat load}
\end{array}
$$

This figure is the heat load or heat requirement for the sample house, expressed in Btu's per Degree Day, and it is the figure used to estimate solar performance.

To find the ratio of heat load to glazing, divide the building heat load by the glazing area.

$$\text{Glazing load ratio} = \frac{\text{heat load}}{\text{glazing area}}$$

$$\text{GLR} = \frac{743}{250} = 2.97$$

The accompanying maps show the glazing load ratio (GLR) necessary in order to achieve a 70%, 80%, or 90% solar contribution to annual heating requirements. For Chicago, they show that a GLR of about 12 will result in a 70% solar contribution, a GLR of about 8 will mean an 80% contribution, and a GLR of about 3 will mean a solar contribution of 90%. The smaller the GLR, the larger the glazing area required. Since 2.97 is less than 3, we know that our assumption of using 250 square feet of solar glazing will give us slightly more than 90% of the annual heat requirement.

The solar contribution can now be applied to the total annual heat load (which is based on the adjusted 4,007 Btu/DD heat loss):

$$
\begin{array}{r}
24{,}550{,}889 \text{ Btu/yr. heat loss} \\
- 20{,}000{,}000 \text{ Btu/yr. internal gain} \\
\hline
4{,}550{,}889 \text{ Btu/yr. net heat loss}
\end{array}
$$

Ninety percent of 4,550,889 is 4,095,800 Btu/yr. supplied by solar heating, leaving only 455,089 Btu/yr. to be provided by backup heat.

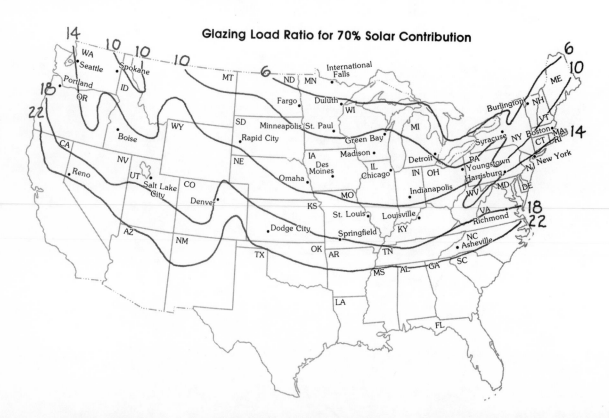

Glazing Load Ratio for 70% Solar Contribution

Glazing Load Ratio for 80% Solar Contribution

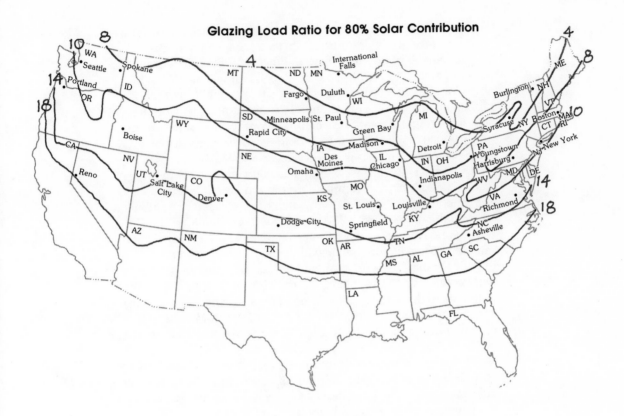

Glazing Load Ratio for 90% Solar Contribution

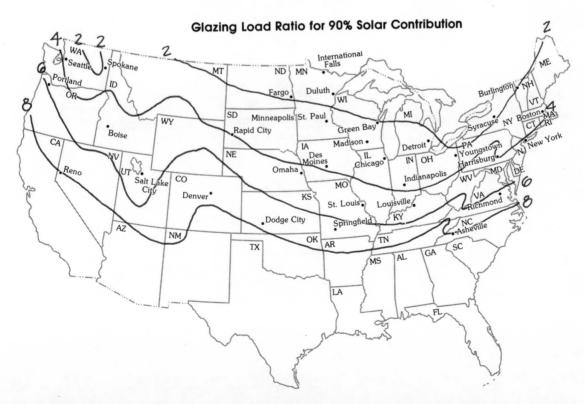

You must be wondering how to work these calculations for a house or location different from what we have given as our sample. Not all of you are going to build a 1,000-square-foot house in a climate identical to Chicago's. Nor will all of you be able to add as much thermal efficiency to a home that you are retrofitting. To make all these numbers and equations work for you, we thought it might be helpful, before we go on to tell you how to build a solar house, to take three other examples and see how certain variations would affect the calculations.

Example 1: Let's start with an easy one and say that you would like to build the sample house in a different location —Minneapolis, for example. How much will the solar contribution be from the glazing area (250 square feet) we computed for the Chicago house? The answer is very simple. The GLR for the sample house will not change with location. It is based on the relationship between heat load and glazing area on a per-Degree-Day basis—not on the *number* of Degree Days. All you have to do is check a GLR map and see what your bottom line will provide you with; thus, a GLR of 2.97 in Minneapolis will earn you a solar contribution of 80% to 90%, rather than 90% or better as in Chicago.

The problem becomes more complex if you ask: How can I get 90% or better in Minneapolis? Yes, of course this means enlarging the glazing area. But how much? One way to get an idea is to use the equation we have used before:

$$\text{GLR} = \frac{\text{heat load}}{\text{glazing area}}$$

But in another form:

$$\text{Glazing area} = \frac{\text{heat load}}{\text{GLR}}$$

We know what the heat load was (743 Btu/DD), and to achieve a solar contribution of 90% the GLR map says there must be a GLR of 2. Therefore,

$$\text{Glazing area} = \frac{743}{2} = 372 \text{ sq. ft.}$$

Yet it is not that easy. You must remember that when you add glazing area to a house, the heat load will change for purposes of making solar calculations. Thus, if you are to use 372 square feet of solar glazing in the sample house, you recalculate the heat load as 673 Btu/DD, and the equation will look like this:

$$\text{GLR} = \frac{673}{372} = 1.8$$

With this new GLR, your house will be getting over 90% solar contribution. If you are really intent on hitting 90% exactly, you'll have to assume a glazing area somewhere between 250 and 372.

At this point, we should stress that we are dealing with averages and variable numbers. Weather changes from year to year, never staying at the average; systems, weatherproofing, and insulation vary slightly in operation, according to how well they are built or installed. Therefore, to say you will get 90% or 80%, or any finite number, is only a relative prediction. We're safe in saying that our sample house in Chicago will get over 90% solar contribution with 250 square feet of solar glazing, and that if you increase the glazing to 372 square feet, you can expect to get the same in the Minneapolis area.

Finally, we should add that design considerations play into the calculations. The use of 360 square feet might work into our plans better than 372, and in fact may be all the area we can get. There is no sense in calling for a glazing area larger than the south face of the house.

Example 2: Let us assume that you own a house identical to the sample house, in the same location (Chicago), but that the house has a north-south orientation rather than east-west. Instead of having a 50-foot south wall, you have only a 20-foot wall. If you added solar glazing to this entire wall (160 square feet), what rate of solar contribution could you expect?

To make the calculations simpler, let's begin by assuming that the window area of the house will remain the same as that of the sample house—that the wall to which you are adding the glazing was previously a windowless wall. To recalculate your heat loss you take the entire heat load we first calculated (4,718 Btuh) and deduct 250 Btuh, which is the gain for a 160-square-foot wall (.024 × 160 × 65°F). Add infiltration heat loss (7,020 Btuh), convert to Btu's per Degree Day (4,242 Btu/DD), and deduct the internal gain (3,264 Btu/DD), to achieve a heat load of 978 Btu/DD. The final equation looks like this:

$$\text{GLR} = \frac{978}{160} = 6.1$$

Now simply check the map and note that for Chicago, a GLR of 8 gives 80% contribution, and a GLR of 3 gives 90%; a GLR of 6.1 would thus yield approximately 84% solar contribution.

Example 3: As a final example, let us assume you own a 20-by-50-foot house in the Chicago area with an east-west orientation. Your house is not as thermally efficient as it might be, but either the house is hard to add insulation to, or you question our premise that you must put in more insulation before adding solar heating. You'd like to see how much solar glazing you would have to add to the south wall to achieve a 70% solar contribution, using only the existing insulation.

First off, you have to calculate your conduction heat loss:

	U	×	Area	×	ΔT	=	Heat loss
Walls (3½ in. insulation)	.069	×	978	×	65	=	4,386 Btuh
Ceiling (6 in. insulation)	.046	×	1,000	×	65	=	2,990 Btuh
Floor (no insulation)	.450	×	1,000	×	30	=	13,500 Btuh
Windows with storm	.580	×	100	×	65	=	3,770 Btuh
Doors with storm	.270	×	42	×	65	=	737 Btuh
					TOTAL		25,383 Btuh

Your infiltration heat loss will go up, too, as you will have a complete change of air each hour rather than a three-quarters change per hour:

$$\text{Infiltration heat loss} = .018 \times 8,000 \times 65 = 9,360 \text{ Btuh}$$

Your total heat loss is 34,743 Btuh, which converts into 12,828 Btu/DD. Deduct for internal gain (3,264 Btu/DD), and your net heat load is 9,564 Btu/DD. To get a rough idea of how much glazing you will need, plug this number into the equation:

$$\text{Glazing area} = \frac{9,564}{12} = 797 \text{ sq. ft.}$$

This square footage is high, because you have not deducted for the gain from solar glazing. Yet it gives a rough idea of how much glazing you would need to add to the house to achieve the modest solar contribution of 70%. With this amount of solar glazing, the roofline of the house would have to change, or you would have to use a solar attic in addition to a glazed south wall. Rather than going to this extreme, you might consider improving some of your insulation first, to see where this would get you.

	Existing house	Improved to
Ceiling	6 in. insulation	12 in. insulation
Floor	no insulation	9 in. insulation
Windows	storm windows	triple-glazed with thermal shutters

You can recalculate your heat loss as follows:

	U	×	Area	×	ΔT	=	Heat loss
Walls	.069 ×		978	× 65		=	4,386 Btuh
Ceiling	.025 ×		1,000	× 65		=	1,625 Btuh
Floor	.032 ×		1,000	× 30		=	960 Btuh
Windows	.100 ×		100	× 65		=	650 Btuh
Doors	.270 ×		42	× 65		=	737 Btuh
					TOTAL		8,358 Btuh

We'll assume that the upgraded insulation and the addition of thermal shutters will cut your infiltration heat loss back to a three-quarter air change per hour, or 7,020 Btuh. Add this to your conduction heat loss for a total heat loss of 15,378 Btuh, or 5,678 Btu/DD. Deduct internal gain (3,264 Btu/DD), and your new net heat loss is 2,414 Btu/DD. Now plug the new net heat loss into the equation:

$$\text{Glazing area} = \frac{2,414}{12} = 201 \text{ sq. ft.}$$

Next, go back and deduct 200 square feet of solar glazing from your heat loss total:

	U	×	Area	×	ΔT	=	Heat loss
Windows	.100 ×		60	× 65		=	390 Btuh
Walls	.069 ×		140	× 65		=	628 Btuh
					TOTAL		1,018 Btuh

Deduct this number (1,018) from your heat loss (8,358), add infiltration heat loss (7,020), convert to Btu/DD (5,287), deduct the internal gain (3,264), and your new net heat loss is 2,023 Btu/DD. When this figure is worked into the equation, you find:

$$\text{GLR} = \frac{2,023}{200 \text{ sq. ft.}} = 10$$

A GLR of 10 for Chicago is less than 12, and therefore more than a 70% contribution. The difference between 750 square feet of glazing and 200 square feet has to be weighed against the cost of adding insulation and thermal shutters. Perhaps you can earn a tax credit for the insulation and shutters, and less structural work would have to be done to add 200 square feet of glazing rather than 750. We think you will agree with us that the extra insulation and thermal shutters are the way to go, especially when you also consider that these strategies will greatly reduce heat gain in summer, keeping your house cooler and more comfortable.

Designing
Your Solar House

5

Now that you have the basic information you need for the design of your solar house, you have to decide who will design it. You can have an architect do it. You can do it yourself. In between, there are variations on the theme—such as purchasing a standard plan to be adapted to your needs, selecting a contractor who will draw up a plan, or attending a workshop/seminar led by an architect.

Anyone who participates in the design of a solar house should have a thorough understanding of solar processes. An architect, a designer, or a contractor who does not have the necessary technical background simply won't do.

Why would you hire an architect? An architect is trained to design buildings, to make them safe—structurally and environmentally— and is licensed to protect health, safety, and welfare. In some cities and counties, an architect's drawings are required in order to get a building permit. And, more and more, architects are trained for the kind of energy-conscious solar design that we have described. In addition, an architect can translate your preferences and ideas into a house that will reflect your needs, life-style, and budget.

If you want to build a well-designed, workable solar house, we think that you should hire an architect, one who understands the concepts of thermal efficiency and climate response and who designs solar houses.

It makes sense to us for you to get all the help you can from the very beginning. This may prevent you from being one of the enthusiasts who are re-inventing some old ideas, such as domes, log cabins, and post-and-beam construction. Domes are expensive, and

120

Detail of the solar-heated
Jacobs House, Madison,
Wisconsin (1948), designed
by Frank Lloyd Wright.

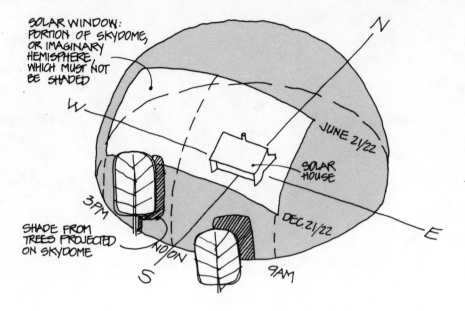

SOLAR WINDOW: PORTION OF SKYDOME, OR IMAGINARY HEMISPHERE, WHICH MUST NOT BE SHADED

SHADE FROM TREES PROJECTED ON SKYDOME

N

W

S

E

JUNE 21/22

SOLAR HOUSE

DEC. 21/22

3PM

NOON

9AM

the spaces within them are awkward and difficult to arrange. Post-and-beam construction is also expensive, compared with stud construction, because of the additional materials required to skin the building. Log cabins are not thermally efficient and are difficult to insulate properly, compared with stud walls.

An architect will help you to evaluate all of these ideas and will select an appropriate building type and construction method for your house, in your climate. He can help to focus your building plans, control your budget, and refine your house. However, if you choose to be your own designer, we hope that the following pages will help you.

In our seminars, typically some participants intend to design their own houses, and others want enough information so that they will understand what their architects are suggesting. We try to give both groups the help they need. In this chapter, our goal is to do the same. We are assuming that the do-it-yourself designer has some knowledge of construction and of architectural drawing.

If you choose to use an architect, here is the way to start. Meet with the architect(s) you are considering, review projects he or she has designed for other clients, and describe your needs. You may see slides or photos of the architect's work, and you may wish to talk with other clients and visit some of their buildings.

When you select your architect, be sure to discuss financial questions, such as building costs, his or her fees, and any other costs that you might expect. Be as clear as you can about your building budget, and outline what you expect to be included. Your architect

Opposite page and this page: Two representations of the solar window, the part of the sky that must remain unobstructed to provide clear access to the sun.

needs to know if you have planned the budget to cover construction costs, land, fees, site work, and landscaping. If it is a building budget only, you will need to think about your other expenses—for the architect's fee, for permits, for water supply, sanitary services (sewer hookups or septic system), utility hookups, and landscaping.

The architectural fee normally is 10% to 15% of the cost of construction for all services through working drawings, paid at intervals throughout the project. An additional 2% generally is charged for assistance in bidding, on-site observation, and contract administration during construction of the building. If you live at a distance from your architect (100 miles or more), you may or may not wish to limit the latter phase of services or to compensate for the additional travel required.

After all of these matters have been discussed, some form of written agreement should be prepared and signed by the architect and client. The agreement will delineate the services to be provided by the architect, the fee established, and the payment schedule. You can use a standard form of contract available from the American Institute of Architects, or just a simply worded letter. A contract is only as good as the trust relationship between client and architect. No amount of legal language will mold that relationship.

Selecting the Site

Designing your passive solar house begins when you choose the land where you will build it. A clear "solar window" is the most important site requirement. This means that between 9 am and 3 pm at the winter solstice, the sun should strike the south face of the house all the way to the floor line.

Where can you find such a site? At this stage in the growth of the solar age, we are finding it easiest to discover suitable sites in rural areas and small communities, where low population densities and large lots usually allow for clear solar access.

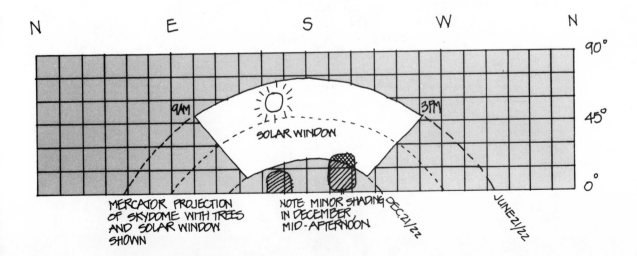

N   E   S   W   N

90°

45°

0°

9AM   SOLAR WINDOW   3PM

MERCATOR PROJECTION OF SKYDOME WITH TREES AND SOLAR WINDOW SHOWN

NOTE MINOR SHADING IN DECEMBER, MID-AFTERNOON

DEC.21/22

JUNE21/22

Here is what you will discover when you look at vacant lots in the city: Since most cities and villages are laid out in a grid pattern, each street will have lots facing toward two opposing points of the compass. If blocks have a long east-west orientation, the lots will face north or south. These are good potential sites for solar houses. Lots that face east and west may not be suitable properties for a solar house, since a house or trees on the next lot south could be close enough to shade the building site.

In subdivisions, site selection is more complex because of the curvilinear streets and odd-shaped lots, where orientation could be in any direction except straight up. Many of these lots have no solar access at all and are totally unsuitable for solar houses.

It is possible to plan subdivisions so that all lots have clear access to the sun. We are working on such projects now, and it is our belief that solar access at building sites should be a basic guarantee, protected by zoning legislation. At the present time very few communities protect your "solar rights," that is, your right to a clear solar window, but we believe that as the demand for solar houses continues to increase, protections will follow—or nonsolar lots will stay vacant. Until ordinances and codes protect you, however, be wary. Find out what can be built on the property south of the lot you are considering, and check allowable building heights and setback requirements for houses and accessory buildings.

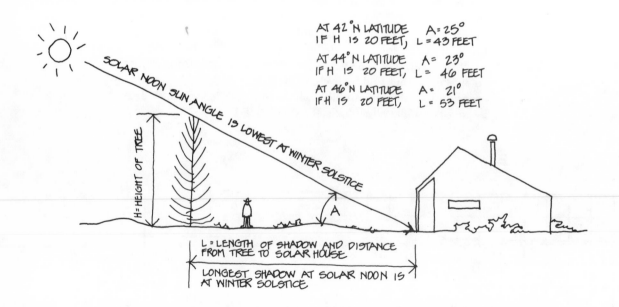

AT 42° N LATITUDE   A = 25°
IF H IS 20 FEET,   L = 43 FEET

AT 44° N LATITUDE   A = 23°
IF H IS 20 FEET,   L = 46 FEET

AT 46° N LATITUDE   A = 21°
IF H IS 20 FEET,   L = 53 FEET

SOLAR NOON SUN ANGLE IS LOWEST AT WINTER SOLSTICE

H = HEIGHT OF TREE

A

L = LENGTH OF SHADOW AND DISTANCE FROM TREE TO SOLAR HOUSE

LONGEST SHADOW AT SOLAR NOON IS AT WINTER SOLSTICE

A treeless site—flat or on a south-facing slope—is a simple choice for a passive solar house, and certainly the easiest site to evaluate. What should you look for? As you search for the property of your dreams, carry a compass. Keep your eye out for a lot that will permit your house to face due south. A slight variation from due south—a few degrees—can be tolerated, if necessary. (Although there will be some lessening of solar gain, this can be compensated for in design by enlarging the collection surface.) Look for any barrier to solar access—buildings, trees, walls—south of your lot.

At the winter solstice, the angle of the noontime sun is lowest (25° at Chicago's 42° north latitude, for instance). This is the critical angle, creating the longest horizontal shadow that you will see each year. A 60-foot-tall oak tree, for example, will cast a 225-foot shadow at that time. To keep your house from being shaded by an obstruction, your building site must be situated a distance of at least 3.75 times the height of the obstruction from that barrier. This distance is calculated to permit direct sunlight to strike the floor line at the south face of the building at noon on the winter solstice.

You or your architect will have to calculate these distances for all obstructions south of the site, from east to west, making a drawing that will project the shadow of any existing or potential obstruction and will tell you where and when the sun will strike your house. If buildings are already on the crucial lot, ascertain their height. Estimate the height of existing trees or measure them with instruments. Your architect can assist here, using instruments or simple geometry. Where construction has not yet taken place on adjacent property, you should assume the most extreme possibility. If you assume that a house is built to the north setback line and that it is as tall as permitted, you will be prepared for the most extreme shading from that potential building. It may be wise to assume that your neighbor will plant some trees at or near the lot line; then you will not be surprised when trees someday grow tall to the south and block the sun.

In selecting an existing house for retrofit to solar heating, it may be difficult to find a building that has a clear solar window to the floor line, especially on urban and suburban sites. If you can find a house with clear access, you are in luck. Sometimes other buildings are so close to a house that the sun does not strike below the second floor, or the roof. In such cases, don't give up. Perhaps a rooftop system will work. You should take care in evaluating a house for retrofit. You can't solar heat a building if there is a high rise south of it.

Since you are unlikely to find a thermally efficient house, the best choice may be a house that needs lots of work. When a house

Opposite page:
Calculating a tree's shadow at noon on the winter solstice.

This page:
A member of The Hawkweed Group uses an instrument to calculate the solar window on site.

undergoes major remodeling, it is easy to add insulation, double-glazed windows, and other thermal improvements.

Even if you plan to do the work yourself, we think it is a good idea to ask an architect to evaluate the house. Hidden structural and environmental problems have a way of coming out of hiding later, and may increase your rehabilitation costs. After that, local codes permitting, you can be on your own and on your way.

As you evaluate properties, don't overlook the usual considerations in site selection. Look for good topsoil and drainage, for healthful surroundings in an area that is close to job, schools, shopping, and so on. If the site is rural, be sure that good drinking water is available, and that soils are suitable for septic systems.

The Newcomer House

The Newcomer House of Bryan, Ohio (opposite page, left), was designed to gather solar radiation over the tops of the surrounding trees. Therefore an extensive solar roof was used, along with direct-gain windows on the second floor (opposite page, right), which also supply natural daylighting.

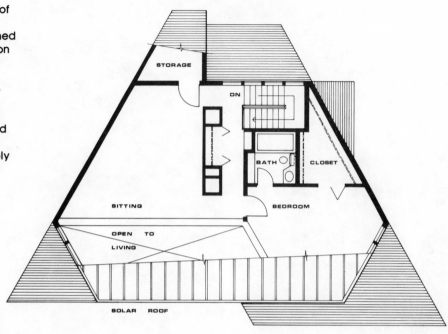

Second Floor Plan

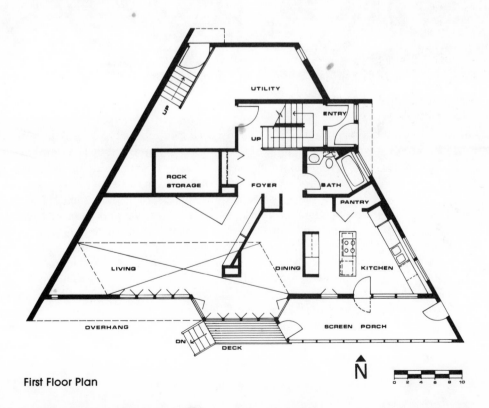

First Floor Plan

Planning the Building Program

When the land has been purchased, you need to pull together the necessary preliminary information. Your architect, if you have one, will ask you to provide a plat of survey for your building lot. This is a measured drawing of your property, made by a registered land surveyor, which should show the location of all utilities and easements, topography at 2-foot intervals, and all trees (over 10 inches in diameter) that are near the building site, as well as other important physical features, such as ponds, streams, and existing buildings. This information is critical in accurate placement of the building on the site. The plat may also be needed for a building permit, and it will be useful even if you are building the house yourself.

The architect or you should check local codes and ordinances to

The Peterson House

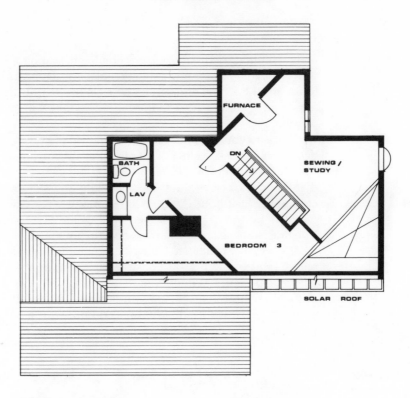

Second Floor Plan

The Peterson House of Deerfield, Illinois, began as a typical tract house, onto which the owner wished to build a solar-heated addition. The two-story addition was designed so that the owner could build it himself. Since the house is located in the woods, emphasis was placed on a solar roof collection system, with direct-gain windows on the first floor.

First Floor Plan

be certain that the house, when completed, will be in compliance with all building codes and zoning regulations. Don't be embarrassed to ask local building and zoning officials for help. It is their job to assist you in complying with the rules, and they will be as helpful as possible.

You and your architect should then visit the property in order to decide the exact placement for your house. If you have trees shading the site, you may want to go through a full year before building so that you can observe the path of sunlight and shadow and can choose the best house site, saving as many trees as possible.

Next, you (and your architect) will define the building program. You must know exactly what you need in your house—number and kinds of rooms, room sizes, privacy requirements—all of your plans and dreams for your solar house. Since your life-style should influence design, it is important to talk about this with your architect or think about it yourself. Do you entertain frequently? Do you preserve your own food? What part do music, the outdoors, gardening, or other interests play in your life?

Solar houses reflect the life-styles of their owners. This page: The Crafts of Evanston, Illinois, included a hot tub when they added a sunspace to their house.
Opposite page: The Sniders of DeKalb, Illinois, wanted a house in tune with nature. With judicious opening of vents and windows, they can get a complete change of air in ten minutes. Their house combines a solar roof, direct-gain windows, and a greenhouse.

The Snider House

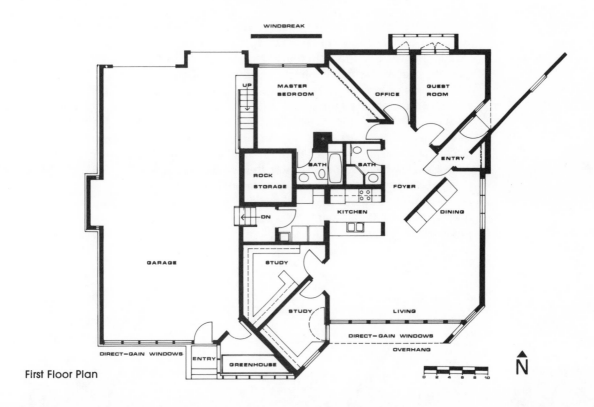

WINDBREAK

UP

MASTER BEDROOM

OFFICE

GUEST ROOM

ROCK STORAGE

BATH

BATH

ENTRY

FOYER

DN

KITCHEN

DINING

GARAGE

STUDY

STUDY

LIVING

DIRECT-GAIN WINDOWS

ENTRY

GREENHOUSE

DIRECT-GAIN WINDOWS

OVERHANG

0 2 4 6 8 10

N̂

First Floor Plan

You may have to modify your plans to keep the total square footage of the house within your budget price, and this is normal. Building prices are established in the marketplace; they are not controlled by either the architect or the owner. Contractors' bids, based upon completed working drawings, will determine the cost of the house per square foot.

This is probably as good a place as any to talk about costs. In the years that we have been designing solar houses, we have had plenty of chances to examine prices for them, and in our experience, solar houses do not have to cost any more than conventional ones. If your solar-heating system is an add-on component or relies on the high-technology approach, it will, indeed, increase the cost of the house. It is for this and many other reasons that we have opted for passive solar houses, in which the heating system is an integral part

The Blount House

The Blount House of Palos Park, Illinois, collects sunlight through two solar roofs. The greenhouse receives ample sunlight in winter once the trees drop their leaves, and the same trees cool the greenhouse in summer.

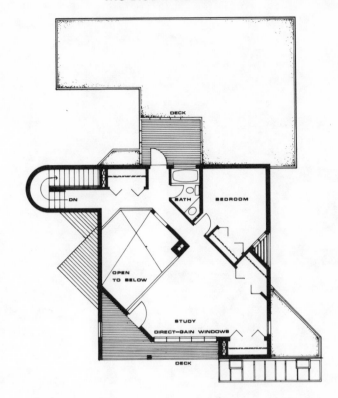

Second Floor Plan

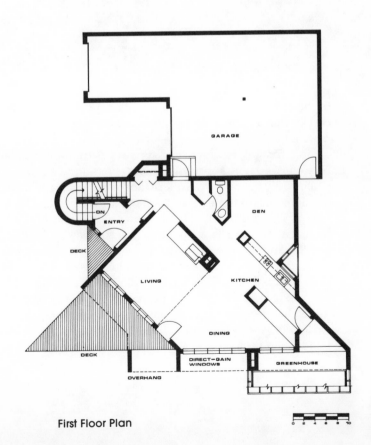

First Floor Plan

of the structure. The building itself is the collection system, with certain readily available materials merely replacing other possible materials.

In Chicago, as we write, a typical custom-built residence costs between $40 and $45 a square foot. For a very elegant house, you could easily spend $60 a square foot or more. The buildings that we are doing cost the same and sometimes less. In fact, at Soldiers Grove, Wisconsin, prices for three of our buildings came in below the budget at $25 to $38 a square foot, testimony to the fact that solar heating and thermal efficiency need not be prohibitive in cost.

Retrofitting an existing house will cost more per square foot, since it ordinarily involves remodeling and addition work. These complex projects, small and personal in scope, frequently call for more changes during construction and more detail than a straightforward new house. The higher cost is also typical of nonsolar remodeling projects.

The Gerut House

The Gerut House of York Center, Lombard, Illinois, was designed to fit into an existing cooperative community with an architectural review board. It is heated by a solar roof in conjunction with direct-gain windows.

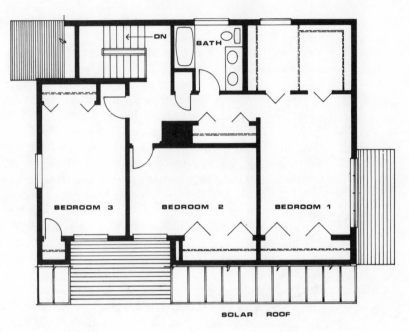

Second Floor Plan

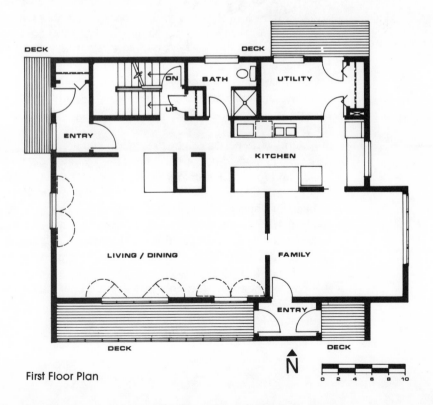

First Floor Plan

The architect will work within the framework of current typical costs, and will try to guide your decisions toward the achievement of your budget goal. If you do not have an architect, you should ask contractors, lumberyards, and builders' associations about current square-foot costs for houses in your area, and allow for inflation.

Many people, when they are planning a home, want to build a great deal of space and include luxury finishes and fixtures, but we believe this is the wrong direction. Think small. Build as little space as you can get along with. Use less-expensive finishes. There are good reasons for shrinking your "want" list. The building will cost less. You will have less space to manage and maintain. Less energy will be needed to heat and light the smaller house, and less energy will have been used to build it. Your architect can help you plan for multiple use of spaces. Building small will be especially useful if you are the builder. The days and hours of labor will be shorter and the beautiful frustration that is part of every building experience will be over sooner.

Don't build what you don't need. A car has a roof on it—do you need a garage? Friends and family always gather in the family room. Do you need a living room, a dining room? Every square foot of a building costs money. Not only the living area, but also basement space, the garage, a loft—even closets and thicknesses of walls are part of the square footage you will be buying. For example, a 2,000-square-foot house with a full basement and a two-car garage may be a 4,600-square-foot building project. Although some parts of the house may be cheaper per square foot than others, all spaces must be included when figuring the cost of construction.

Be realistic, then, about your house. Let it suit *your* life and serve *your* needs. Your architect can translate those needs into a suitable passive solar house. Be firm with yourself. Cut your "want" list to the bone.

Before your house can be designed, your architect (or you) must assemble climatic data for the area. The *Climatic Atlas of the United States* and *Local Climatological Data* summaries are available from the National Climatic Center in Asheville, North Carolina. ASHRAE publications also contain useful information. Since radiation data are not gathered in every city, some interpolation may be necessary between the nearest data stations. This requires skilled judgment, but with the help of local climate information, it is possible to come up with satisfactory figures.

Your architect should now prepare site drawings, showing heating requirements (expressed in Degree Days), latitude, azimuth of the sun at equinox and solstice, altitude of the sun, wind patterns, and humidity. Mesoclimatic effects should also be charted. Cold air drains to the lowest point in a valley, snow showers are heavy in some areas—these and other effects will influence design decisions.

If you are going to design your house, you will have to gather the climate information yourself and learn to use it, as discussed earlier in this book. We have listed some helpful books in the Bibliography that will give you more information about climate.

Now it is time to design your solar house.

Designing Your House

Your architect usually will start with tiny quick sketches, to examine various possible space configurations. As he or she designs the house, the architect will consider the site, the climate requirements, your building program and budget, the structural and mechanical systems of the house, and the harmony and balance that will help to give you a handsome building. If you are your own designer, you must do the same.

Standing outdoors in the wintertime, you turn your back to the wind. So should your building face away from the winter wind,

The north side of the Snider House shows two ways of protecting windows from harsh winter winds.

which usually comes from the north or northwest. This means no north windows; it means earth integration and the use of windbreaks.

Is there a view? If your site has an outstanding north view, you may be tempted to insist on a design that opens up on that view, even though this will violate the thermal integrity of the building. Windows on the north will never receive solar gain and will only lose heat on this winter-windy side of the house; but maybe you could compromise, with a viewing room or porch to be closed off during cold months, or a small viewing window that is thermally shuttered when not in use. On the other hand, perhaps there is a pseudo-colonial house, gravel pit, or other eyesore to the south that needs to be screened from sight.

Ask yourself how you can use the sun. Can you bring it directly into the living space? If all living areas are on the south, each room can be directly heated. Utility spaces—storage, stairs, halls, and closets—can be situated on the north to serve as buffers to the living spaces.

Relationships between rooms must be analyzed for both climate response and program needs. The latter is typical in any building program, solar or not; the former should be, also. The site may dictate a two-story house. If so, bedrooms could be on the second floor receiving the last warmth from the wood-burning stove at night. However, bedrooms on the first floor will remain cooler, with heat rising to the living spaces above. The choice depends upon the climate, the heating methods, and individual preference. In an area of hot summers, first-floor bedrooms may be more comfortable. In a cold climate, you might want the additional warmth in the bedrooms at night.

To get sunshine in every room, you will want a long east-west plan. The kitchen could be at the east. It's nice to have breakfast—and prepare it—in the morning sunlight. Especially in winter, it lifts the spirit. Bedrooms on the west would be fine, but avoid west windows, which would heat the rooms on late summer afternoons. Let these rooms instead open to the south. Overhangs on the south will protect all rooms from the high summer sun.

How shall the sun be used to heat this building? The floor or the walls could be used for heat storage. Or you could use a mass heat wall, but this would reduce the window area. Perhaps a solar attic is best for this house. Or maybe a sunspace or a greenhouse. They are all superb solar heaters. Do you want to manage the system manually—manipulating dampers, shutters, and vents? Most people who want a passive house are prepared for these

The north windows of the Snider House are protected by a third layer of glazing (this page) and by a free-standing wall (opposite page, top) to deflect winter winds.

solar chores. However, if no one is home at critical times, perhaps some automatic controls will be needed.

Thermal efficiency is a thread running through all of this exploration. Earth berms, thermal shutters, insulation, triple glazing, air-lock entries—all of these are the elements used to keep in heat.

Renewable materials are appropriate for an energy-conserving house. Maximize the use of wood. And try to utilize regionally available materials: By using less energy for transporting materials, you can reduce building costs. Always keep building costs in mind. Keep the house simple. Keep it small. Keep it within the budget.

The building will be drawn up in preliminary sketches, showing floor plans, elevations, perhaps a section through the building, and a site plan. With its long east-west plan, all the living spaces will be directly sun-heated and naturally lighted. The house could be earth integrated, at least with a berm at the north wall. In a rural area or small town, backup heat could be provided by an airtight wood stove.

If you are your own designer, do not skip this preliminary stage. It is the time to test your plan, to be certain it is right for you. The architect will go through many schemes before accepting one, and so must you. We like to make a simple cardboard study model, to help us to develop the building and to help our client understand it more easily. Your architect will meet with you to

Below: Morning sunlight warms the kitchen of the Catoni House, McHenry County, Illinois.

Cardboard model of the
Gallimore House.

review climatic data and drawings. After you have examined the
total concept, you will see the model as well.

At this point, the design is still fluid and easy to change. Your
meeting or your personal review of your design should elicit the
desired changes. A closet, a sliding door, privacy for the master
bedroom, more pantry space—now is the time to incorporate
these ideas. But remember the budget! Many clients say, "If we're
building *this*, we might as well add that." Avoid "might as wells."

Take time to study the sketches carefully. It is easier and
cheaper to change lines on paper than to change studs and
foundations. You may need more detailed information—a
perspective of a kitchen or a section through a loft space—to
clarify the design.

At last, it will all come together. The sketches will reflect your
needs and the demands of the climate. If you are designing your
own house you will probably not make further drawings, but will
now take your sketches to a contractor. He will work out the
construction methods and materials with you and give you a price.

Your architect, when you have approved the design, will
produce working drawings and specifications for your house.
Working drawings are scale drawings, with sufficient detail for
construction. Written specifications concerning materials
accompany the drawings, which will include site plan, floor plans,
elevations, cross sections, wall sections, electrical and mechanical
plans. Larger scale drawings, called details, will provide more

information as needed. An architect may work with consulting engineers for structural, mechanical, and electrical design, or may have these capabilities on his or her own staff.

You will have one or two progress meetings while the drawings are being made. During these meetings you may still make comments and make certain that the house is appropriate for you in all ways.

Finally you will select finishes for the flooring, walls, and doors and put your personal imprint on the house with choices of fixtures, siding, and the like. When the drawings are finished and approved, it is time to build.

Preliminary plans, the Gallimore House, showing south, east, and west elevations

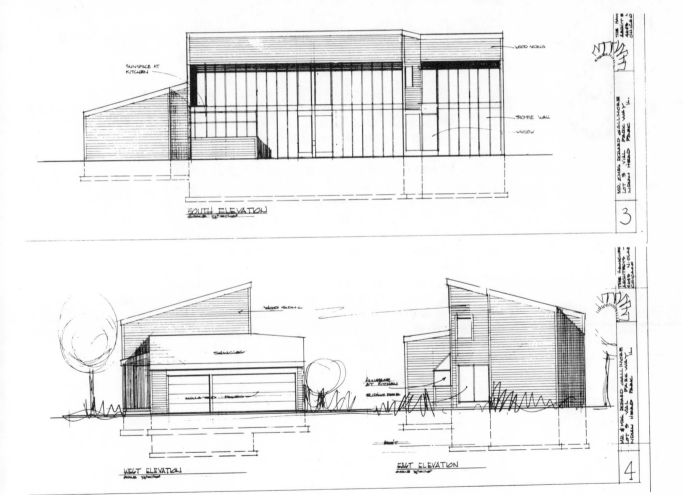

Nailing
It Together

6

With working drawings and specifications completed, you are ready for the next step—construction of the house. You may wish to choose contractors to build your house, you may decide to build it yourself, or you may choose to share construction with selected contractors.

The 14 trades involved in the normal building process are excavation; *concrete; masonry; carpentry;* millwork; *glazing and windows; roofing;* drywall; *insulation;* the mechanical trades of *heating, electrical, and plumbing;* as well as ceramic tile and *painting.* You will notice that there is not a solar contractor. Of these trades, the ten in italics may be involved in the solar portion of the job.

As expenses and interest rates increase, cutting your costs by doing all or part of the construction yourself will become more and more important. The process is time-consuming, and we guarantee that your leisure hours will disappear. Weekends, evenings, and vacation time will be devoted to building your house—not just for a few weeks, but for a year or longer. If you have it in you to do your own construction, plunge in. There is an enormous satisfaction to living in a home that you built yourself. You know all of its inner secret workings—each bolt, nail, and foot of insulation. And the money you save will be in your pocket.

If you have an architect, you should tell him or her what you plan to do and ask for guidance. The architect can help you through the tough spots, and can tailor construction techniques for your level of ability.

We have designed several owner-built houses. Owner

142

The second floor of a
house is framed
and braced.

The Remus House

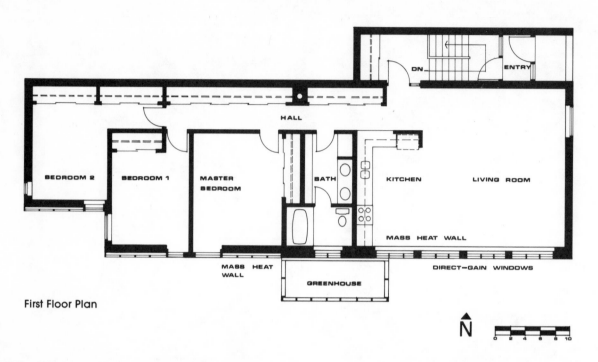

BEDROOM 2 **BEDROOM 1** **MASTER BEDROOM** **HALL** **BATH** **KITCHEN** **LIVING ROOM**

DN **ENTRY**

MASS HEAT WALL

MASS HEAT WALL

GREENHOUSE

DIRECT-GAIN WINDOWS

First Floor Plan

N

0 2 4 6 8 10

The Remus House, Hixton, Wisconsin. The couple building this house hired a contractor to pour the walls and are doing the rest of the work themselves, including the repair of an old windmill to pump their water.

participation has varied from total construction to management and coordination of the work of contractors. Some clients choose to do only some of the finishing work—painting, tiling, and the like. On the other hand, a couple near Hixton, Wisconsin, hired a concrete contractor to pour the walls of their earth-integrated house and are doing the rest of the work themselves. They repaired an old windmill to pump their water. They cut all of the necessary timber from their land, and brought in a portable sawmill to cut the logs to the necessary dimensions.

Another client is using recycled materials, free for the taking. He demolished an old building and salvaged enough material from it to build his new house.

Lack of home-building experience didn't daunt a school counselor who took a year from his work in order to build a home for his family. With the hired aid of his carpenter brother-in-law, he constructed a handsome house, finished with exotic wood paneling and imaginative loft spaces.

A real-estate agent contracted all of the construction work, but he personally washed and installed the rock for his heat storage to avoid the risk of a careless job.

Intense interest in the solar aspects of his new home led a university professor to build his own site-integrated flat-plate collectors and to balance the system himself.

Another client simply enjoys the entire undertaking, zestfully digging excavations and working along with the tradesmen to build a large and elegant new home.

One couple, an airline pilot and stewardess by profession, are finishing their interiors with the same careful concern they had during the design process. They want perfection, and they achieve it, in part, through their own work.

A truck mechanic and his office-manager wife, camping at their home site on weekends, worked with their carpenter crew on an earth-integrated home. When progress permitted, they moved into the house and completed the job.

Soliciting Bids

If building a house is not your cup of tea, you will want to send copies of the working drawings and specifications to contractors for prices, or to one specific contractor if you wish to negotiate an agreement directly. If you are working from your own sketches, the contractors (or contractor) will review these, make assumptions about materials and finishes, and give you a price. This bidding process will result in precise written prices from the contractors.

You can look for competitive bids in any of three possible combinations. One of your options is to choose a general contractor who will give you a price for doing all of the work required, from excavation through completion of the building. This contractor will coordinate the entire job, managing the work for you and arranging for the proper sequencing of trades. As progress permits, usually monthly, he will request payment for the work completed and the materials incorporated in the building. It is his responsibility to pay workmen and suppliers, as well as subcontractors who have worked on the job, and to provide you with documents ("waivers of lien") stating that all suppliers, contractors, and workmen have been paid in full.

A second possibility is to select a general contractor who will give you a price for everything except the mechanical trades. You will then get separate prices for heating, plumbing, and electrical work. The general contractor will be responsible for all parts of the construction job except the mechanical work. You (or your

The Loos House, Libertyville, Illinois. A school counselor took a year off to build himself a unique solar house.

The Loos House

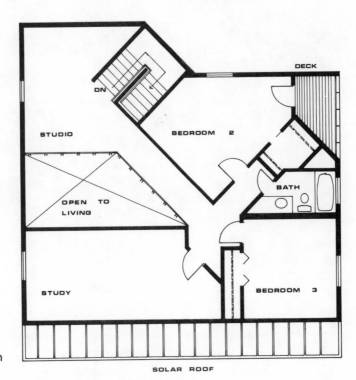

STUDIO

DN

BEDROOM 2

DECK

OPEN TO LIVING

BATH

STUDY

BEDROOM 3

Second Floor Plan

SOLAR ROOF

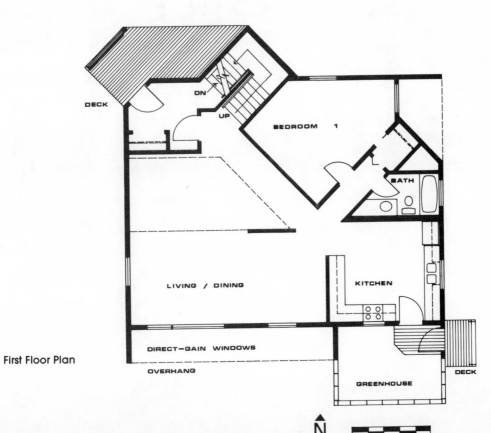

DECK

DN

UP

BEDROOM 1

BATH

LIVING / DINING

KITCHEN

DIRECT-GAIN WINDOWS

First Floor Plan

OVERHANG

GREENHOUSE

DECK

N̂

0 2 4 6 8 10

The Catoni House, with solar roof, direct-gain windows, and greenhouse.

The Catoni House

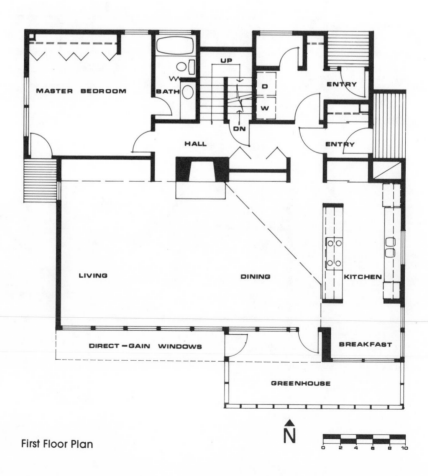

MASTER BEDROOM

BATH

UP

D W

ENTRY

HALL

DN

ENTRY

LIVING

DINING

KITCHEN

DIRECT-GAIN WINDOWS

BREAKFAST

GREENHOUSE

N

0 2 4 6 8 10

First Floor Plan

architect), along with the general contractor, will coordinate the work of the mechanical trades. You will sign separate contracts for the work of each contractor and will pay each directly.

Your third option is to obtain separate bids from contractors representing each trade required in the construction of your house. All 14 trades, then, would be solicited for bids for the job. Coordination of construction would be your responsibility or, if you prefer, that of your architect.

Finding bidders for any of these options is a matter of detective work and perseverance. Check local phone books, ask your architect, talk to people who have recently built a house, note names of contractors at construction sites and on the sides of trucks. When you have put together as many names as you can, the next step is to call all of these contractors to ask if they want to bid on your house.

The Catonis work on finishing the interior of their home, in between their jobs as airline pilot and stewardess.

You will need to supply prints of the drawings and specifications to all interested bidders. A general contractor will want at least three sets of prints. For the second option, you will need to supply three sets to the general contractor and one set to each mechanical contractor who is bidding. If you are taking separate bids, each contractor will need one set.

For each trade or kind of contractor bidding on your house, you will want at least three bids. Good old free enterprise is on your side here. It is such competition that helps to keep prices reasonable. If you are seeking separate bids, this means that you will be giving drawings to at least 42 different contractors. And you will be talking to even more contractors as you look for bidders, answer questions, and cajole contractors into giving you a bid.

It is important to establish a due date, with contractors expected to return their bids by that time. Fifteen to 30 days is sufficient. If you don't set a date, you may never get a price, since contractors operate on the "squeaky wheel" principle. You must call each contractor several times after delivering prints, to be sure he is bidding and that questions are answered.

If any contractor tells you, at this time, that he will not bid, you should get your drawings back and look for someone else. Some contractors will drop out, for a variety of reasons—other work has opened up, there is not time for this project, insufficient cash flow is available.

Sometimes a contractor will tell you that your house is too difficult, that he doesn't understand it and won't give you a price. If this happens, don't be disheartened. You are no doubt fortunate. A contractor who doesn't have the patience and the intelligence to understand your house is not the person to build it. You should be aware that this is common when a contractor is presented with a building different from his accustomed workload. Many homes designed by architects, even nonsolar ones, receive this response. Be patient, and look for more contractors.

A contractor may request more time to turn in his bid, and you may decide to allow this extension. Notify the other bidders, then, that they all have additional time. Be careful, however, not to permit the bid process to be drawn out too long. A serious bidder will get on with his work as rapidly as he can. Encourage speed in bidding, so you can get the building under construction as soon as possible.

If you have reason to think that some of the contractors are simply not working on the project, you may want to look for more bidders. We cannot emphasize enough the necessity of having numerous bids from contractors.

Contractors will have some questions as they develop their bids. Refer these to your architect, if you have one, because he is accustomed to interpreting drawings for contractors.

When the bids come in, you and your architect should check each one to make sure that nothing was omitted. Each bid should include all labor and material required to perform the proposed work according to the drawings and specifications. If the contractor suggests any changes or substitutions, they must be in writing, and you will have to decide whether these changes are acceptable to you.

It is common to have a wide price range among the bids. One bid may be double the price of another—for exactly the same work. Ordinarily, the range of bids will include some that your budget can accommodate. If not, get more bids!

You should select the contractors whose bids and reputations are the most suitable, based upon price and work to be performed. This does not necessarily mean choosing the lowest bidder, since price is only one consideration. Look at the work of each contractor. Talk to former clients. Check credit references. When you are totally satisfied concerning capability, financial responsibility, and prices, you are ready to inform the chosen contractor(s) that he has been selected to build your solar house.

A contract for use in these projects, called the Owner Contractor Agreement, is available from the American Institute of Architects. Developed jointly by the AIA and the Associated General Contractors, it is designed to protect the interests of all parties. You may find this a useful instrument for your agreement with your contractor(s). We suggest discussion of this crucial matter with your architect and with your attorney.

Before the contractor begins his work, you should expect him to give you a Certificate of Insurance, for liability, on this job. As the owner, you should also have insurance coverage for liability, fire, theft, and wind damage. Your insurance agent can advise you.

Solar Building Process

At last, within two weeks after the contracts are signed, you can start building. The process is the same whether work is coordinated by a general contractor, your architect, or you. The following overview points out the aspects of a typical house construction project that should receive particular attention when you are building your solar house.

First, the contractor, the architect, and you will meet at the site to lay out the building. The contractor should be instructed to bring a transit, an instrument used in setting grades and establishing elevations. Another essential instrument to have is a compass, so you can establish due south and site the building facing the sun. Stakes will be set to mark the building perimeter. A marked stake or a nail driven into a tree will indicate the first-floor elevation.

This is also the time to determine what trees must be removed, where topsoil will be piled during construction, and what plants and

Above: A concrete contractor uses a transit to insure a level job. Below: The footings have been poured and the forms for the foundation are ready to be filled at the Pleasant Valley Outdoor Center, an earth-integrated building near Woodstock, Illinois.

trees must be protected. You should not allow trees to block solar access. However, in order to preserve as many trees as possible, you may permit the house to deviate slightly from due south. At the southeast and southwest corners of the building, trees will serve as natural air conditioners in summer.

With backhoe or blade, the first tradesman arrives to excavate for footings and foundations. He starts by stripping the topsoil and stockpiling it for finished grading later. The excavation must closely follow the building layout and should be 2 feet larger all around. The additional 2 feet allows working space to set forms for concrete, to build block walls, to insulate at the outside perimeter of walls, to dampproof, and to put drain tile around footings, if it is needed.

The depth of the excavation should be as shown on the working drawings. A basement is usually 7 feet below grade. For a crawl space, the excavation should go to at least 2 feet below the joist level. Excavation for any below-floor heat storage is done at this time, also.

Dug to halfway through the footing depth, the excavation

Above, left: A concrete contractor tightens up reinforcing rods.
Above, right: The first-floor walls of the Pleasant Valley Outdoor Center are dry and ready for framing to begin.

should be level. Since concrete should be placed on undisturbed soil, overdigging will require additional concrete to be poured—an unnecessary expense.

Soil tests are not usually made in home construction, unless there is reason to believe that the site has been filled with extraneous material. If the condition of the soil at the bottom of the excavation seems soft or unusually wet, you should ask your architect to determine the suitability of the soil to support your building.

With the excavation complete, the concrete contractor is ready to begin. Using 2-by-12's and 2-by-6's, he will form up the footings. He will need a transit to set the bottom of the footings and to insure a level job. Since the big excavation equipment is, by its nature, imprecise, the concrete contractor will have to square up the building to the precise dimensions required. With forms in place and leveled, other trades must place pipes and sleeves, as needed, to go through the footings, and also pipe for drain tile and a sump pump to go under the footings.

Reinforcing rods are not normally required in footings on good soil. A key, or keyway, is cast into the footing. A key is a cast-in-place groove in the top of the concrete footing that

The first-floor walls of this house are framed and braced. Note the openings for direct-gain windows, facing south and southwest.

provides for an interlocking of foundation wall and footing.

The depth of walls and footings extending below the frost line varies with climate. For the north central part of the Midwest, the frost line is 5 feet below the existing grade. The footing depth is 1 foot. Six courses of block plus the footing give a 5-foot frost wall, as does a 4-foot poured concrete wall and a footing.

After at least two days to allow the concrete to dry, the forms are stripped and preparations are made to build the foundation walls. If the walls are to be concrete, forms are erected and reinforcing placed before the walls are poured. For concrete-block walls, where soils are suitable, the mason arrives with block, mortar, sand, and mixer. He and his men lay up the block walls in accordance with the drawings. If a basement rock bed is planned for storage of heat, its walls go up at the same time.

Before backfilling around the foundation, several steps should be taken to protect from moisture and heat loss. The exterior walls of the foundation should be dampproofed with two coats of sprayed-on pitch if there is a basement. If there are serious water problems, a membrane waterproofing may be required. Drain tile is installed around footings now, if needed, with gravel over the tile. Rigid insulation—polystyrene boards—is placed against the exterior walls and held in place with mastic until backfilling secures it permanently. Then it is time to backfill.

Next comes the carpenter, the generalist of tradesmen, preceded by a truckload of lumber. Your solar house should be built entirely of wood, except where other materials are required for the passive solar components, since wood is a renewable resource and a good insulator, with its many tiny cells that trap air.

First, the carpenter places a sill seal over the bolts that have been cast in the concrete to receive the wood sill. This insulation strip reduces infiltration through the space between the concrete or masonry and the wood, and it should be continuous. The wood sill is set in place over the sill seal, bolted to the foundation, and caulked.

If the design requires a beam, it is set now, and then the contractor puts in the 2-by-12 floor joists and the subflooring.

Next, the walls go up. To allow for 12 inches of insulation, exterior walls should be 12 inches deep, made of two layers of 2-by-4 studs at 16 inches on center (that is, 16 inches from the center of one stud to the next), spaced 4 inches apart. Following the drawings, the carpenter leaves openings for doors and windows.

Members of The Hawkweed Group help Bob Selby complete the solar roof of his house. Sheets of black-painted corrugated barn roofing are being nailed in place over black-painted metal pans, which are fitted between the rafters. Air vents (not visible) are at top and bottom. A layer of fiberglass glazing will complete the roof.

Interior partitions follow, with openings left for doors. The partitions will probably be 2-by-4's at 16 inches or 24 inches on center, in accordance with local codes.

If the solar technique is a mass heat wall, the mason will be building it at the same time, leaving openings for vents and ties for glazing. If other solar details are to be built, such as sunspaces, a greenhouse, windows for direct gain, or a solar attic or roof, the carpenter will frame the structure (rafters for roofs, studs, and headers) for future glazing by the glazing contractor.

Once the roof is framed, the enclosure of the house can begin. Exterior siding, windows, roof deck, and shingles—all are put in place now. Everything should be fitted tightly, and the roofing should be done with care to insure a watertight house.

The owner of a solar house staples a vapor barrier over the insulation (top) and does his own electrical wiring (above). Right: A carpenter completes a solar roof by caulking the nail holes on the wood strips that hold the fiberglass glazing in place.

Now come the mechanical trades—the electrician, the plumber, the heating contractor. They will be working along with the carpenter to place conduits, pipes, and ducts.

As the electrical contractor wires the house and roughs in outlets, it is important that outlets in exterior walls be built so they do not allow cold-air infiltration. Each outlet is a break in the exterior insulation of the house and will leak heat to the outdoors unless it is protected. An insulating collar is available for this unique requirement.

The heating contractor will install the backup heating system. If that system is a wood stove, you may wish to do the job yourself. For safety, follow the recommendations of the National Fire Protection Association.

The plumber, in addition to installing the piping and bathtub, will also rough in for the toilet, sink, and lavatory. If you are using a passive solar water heater, he will pipe it.

After the mechanical tradesmen have completed their work, it is insulation time. You should use 12 inches of fiberglass in the walls and 24 inches in the ceiling. This is the time when you should join the construction crew, if at no other. Watch carefully for all of the corners, nooks, and crannies that go unfilled. Every house needs a detail person now—to walk around with hunks of insulation and fill up all of the voids. Once the drywall is up, it's too late. Wear gloves and a mask while insulating to avoid touching and inhaling the material. After the holes are filled, a 4-mil plastic vapor barrier goes on, stapled to the studs.

Drywall goes up next, and the end of the job is near. Many drywallers will object to the use of a vapor barrier. They prefer to glue the drywall in place, and a vapor barrier prevents them from doing this. Nevertheless, you *must* have a vapor barrier, and drywall manufacturers recommend it. As air moves from the warm house through the wall, moisture condenses. The vapor barrier holds this moisture on the warm side. If it moves through to the framing members of the house, the moisture causes rot. Insist on a vapor barrier and have the drywall nailed or screwed in place, as specified.

The house is now ready for all of the finishing processes. Flooring and ceramic tile are put down and trim installed. Kitchen cabinets, doors, and thermal shutters are installed and hardware put in place. Plumbing fixtures are set. The architect or you should make up a "punch list" of all the items to be corrected or completed, for the contractors to check off. When these items are taken care of, moving day is at hand.

The owner acts as detail person, sticking small hunks of insulation in cracks around windows and doors, or in any spot needing extra insulation. He's made one mistake, however: He's forgotten to wear a mask and gloves.

Living
in a Solar House

7

How do you live in a solar house?

First, you become very conscious of the sun—its presence and its intensity, the warmth and brightness that emanate from it. Cloudy days become a challenge. You count their frequency and note their effect on the pattern of heating your house. The quality of each day becomes important, and you find yourself checking each morning—first thing—to see what the sun is offering you today.

The day starts and ends with the tasks you must do to operate your solar house, as is true with all houses. These tasks vary with the type of solar system you have. If you have a direct-gain system, with thermal shutters, each sunny morning you open the shutters and welcome the sun into your house. Each evening you close the shutters so you can hold the heat you have collected.

If you have a mass heat wall with hand-operated vents, you open these vents when the sun shines to draw heated air into the house and you close them again at night.

Your sunspace or greenhouse requires opening and closing shutters and vents, as well. The precise jobs depend upon the design of the space. And with a greenhouse comes the pleasure of spending time each day with your garden plants—watering, cultivating, planting, transplanting, and admiring.

Air-handling systems or remote rock storage beds coupled with solar attics have filters that need to be changed regularly, with the frequency depending upon the amount of dust and dirt attracted. Large dogs, small children, and fluffy cats shorten the time between changes.

158

The exterior of a solar house, showing overhang, direct-gain windows, and ventilation louvers. The shrubs beneath the windows cool air before it reaches the louvers.

Right: A solar greenhouse provides not only welcome heat to your home, but year-round plants and vegetables as well.
Below: Living in a solar house makes you check every morning to see if it will be a sunny day.

You also find yourself, as a solar householder, far more conscious of the need to conserve heat. Your house is designed to be very energy conservative, and you will want to cooperate with this design in several ways. The strategies you use are simply common-sense methods of saving heat, which were familiar to our parents and grandparents.

The windows that are not a part of your solar heating system are also fitted with thermal devices of some sort—thermal shutters, or insulated draperies or shades. These should be closed up tight every night in cold weather to prevent heat loss through the windows. In fact, on cloudy days, if no one is to be at home, you may want to leave these shutters closed all day. If the house is light and bright, perhaps some shutters could be left closed on cloudy days when you are at home—especially if the weather is very cold.

One easy but essential chore for owners of solar houses is closing the thermal shutters at night (above) and opening them again the next morning (right).

Thermostats should be turned down to at least 55° Fahrenheit at night and when no one is at home. The top setting during the day should be 65°.

You will probably have a conventional hot-water heater as a backup to your solar heater. Turn this heater on only when you need hot water and your savings in energy will be significant, with minimal inconvenience.

The air lock that helps to prevent heat loss at the entrance to your house will work very well if all family members close one door before opening the second. We can attest that this does not always happen, and you may need to drill the family on "solar door-opening techniques." Everyone grows up to the refrain

Above, left:
"Closethedoor!"
Above, right: Stoking the
wood stove.

"Closethedoor!" By the way, if you have a large dog, be sure your air lock is large enough to hold both of you while you are "between doors."

If you need to be away for a few days, your heat requirements will be less, since you just want temperatures to stay above freezing. Leave the system shutters or vents open to grab available heat; close the other window shutters and button the house up tightly. If cloudy days occur, your backup system will protect the house, with the thermostat set as low as possible. When you return, you may need to warm things up a bit, but it won't take long.

If you are fortunate enough to have woodlands on your

property, wood is probably one of your backup heating methods. In this case, there is one additional job to do and a pleasant one, we think. This is the selection of trees for cutting, the felling of each tree, cutting and splitting, loading and carrying to the woodshed, unloading and stacking. Then comes a new set of tasks—bringing in the wood, stoking the fire, feeling the glow that penetrates to the bone and warms as no other heat does. Removing the ashes to store for your summer garden is the final chore. A whole world awaits you, as you learn the techniques of cutting firewood and the management of the woodlot, woodpile, and stove. Think what a health club would charge for such good exercise!

As you can see, the demands of a solar house are not difficult or time-consuming. In fact, we find that we enjoy the experience of controlling and working with our houses, rather than depending upon machines or repairmen to manage it all for us. A solar house is simply a comfortable house with even heat—a nice place to live in and come home to.

Opening louver vents to let in some air.

Our
Solar Future

8

By using the strategies that we have described, all new buildings could be heated by the sun, and existing structures could be retrofitted for solar heat. Individual residences—solar-heated and climate-responsive—are building blocks for the communities of the solar future. Two important issues need to be addressed in order to smooth the path for that future: First, our communities need to retain the buildings already standing, whenever possible, and retrofit them for solar heating. Second, communities need to guarantee access to the sun at each building in future developments and in redevelopment projects within cities.

Urban Projects

Rehabilitating Through Retrofit. Let's look at conventional urban housing. Many existing residential buildings are poorly insulated or not insulated at all. They were not designed to conserve energy. Their heating systems, by today's solar standards, are obsolete. But to tear these buildings down and replace them with new, thermally efficient, solar-heated housing, would require an enormous expenditure of energy and money.

We studied a 55-year-old building in poor condition. Like many other buildings in its Chicago neighborhood, it was being allowed to go to ruin. And, like many others in its neighborhood, it could be renovated into a thermally efficient, solar-heated building with greenhouses for tenant use. Restored, it would provide fine homes while conserving energy.

In a project that heralds this kind of redevelopment, a group of

164

Site plan for solar-heated multi-family residences that could be built on one block in any city with a cold climate.

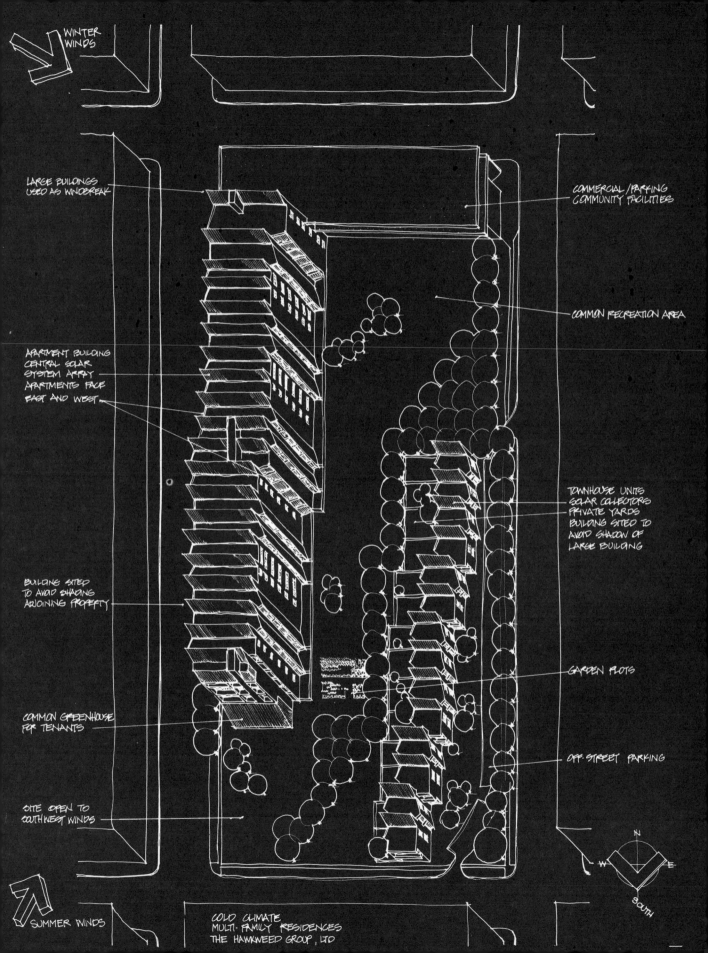

WINTER WINDS

LARGE BUILDINGS USED AS WINDBREAK

COMMERCIAL / PARKING COMMUNITY FACILITIES

COMMON RECREATION AREA

APARTMENT BUILDING CENTRAL SOLAR SYSTEM ARRAY APARTMENTS FACE EAST AND WEST

TOWNHOUSE UNITS SOLAR COLLECTORS PRIVATE YARDS BUILDING SITED TO AVOID SHADOW OF LARGE BUILDING

BUILDING SITED TO AVOID SHADING ADJOINING PROPERTY

GARDEN PLOTS

COMMON GREENHOUSE FOR TENANTS

OFF-STREET PARKING

SITE OPEN TO SOUTH WEST WINDS

SUMMER WINDS

N
W   E
SOUTH

COLD CLIMATE
MULTI-FAMILY RESIDENCES
THE HAWKWEED GROUP, LTD

The residents of this three-story brick apartment building in Chicago are planning to convert it to solar heat, as shown in the diagrams.

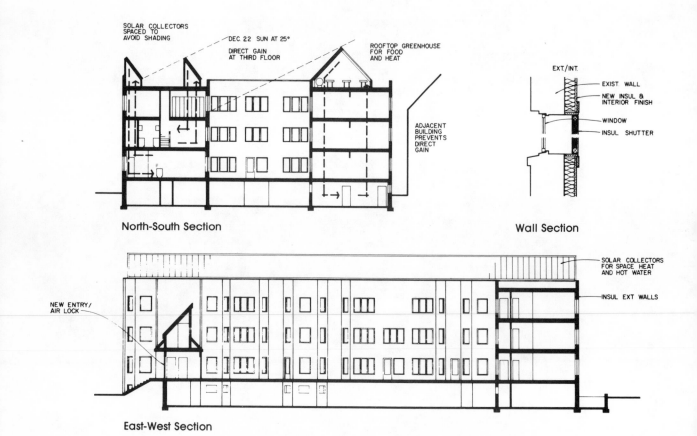

SOLAR COLLECTORS SPACED TO AVOID SHADING

DEC 22 SUN AT 25°

DIRECT GAIN AT THIRD FLOOR

ROOFTOP GREENHOUSE FOR FOOD AND HEAT

ADJACENT BUILDING PREVENTS DIRECT GAIN

North-South Section

EXT/INT.

EXIST. WALL

NEW INSUL & INTERIOR FINISH

WINDOW

INSUL SHUTTER

Wall Section

NEW ENTRY/ AIR LOCK

SOLAR COLLECTORS FOR SPACE HEAT AND HOT WATER

INSUL EXT WALLS

East-West Section

community residents are preparing to rehabilitate a similar building on Chicago's South Side. The 22-unit building, erected at the turn of the century, is still sturdy, but in disrepair. It is a three-story structure, with its apartments arranged around a central courtyard that receives little sun. The walls, of common brick and face brick, are uninsulated, as is the roof. The windows are single-glazed, and the entries have single doors.

To make an older building thermally efficient—the first step in retrofitting—you must insulate the walls and the roof. An easy way to insulate the walls, we have found, is to make them thicker by a process called furring out. Either on the inside or on the outside, an extra 6 inches of space is provided by furring out, and this space is filled with 6-inch fiberglass batts of insulation. A new wall is then built over the insulation. To insulate the roof, there is usually enough space between the ceiling of the top floor and the roof to add 6 to 12 inches of new insulation. A building in very poor condition is a perfect candidate for this kind of treatment. Walls, windows, and doors usually need to be replaced in any case, so furring out and insulating fit nicely in the schedule.

The Chicago building will be insulated at the inside walls and at the roof. All windows will be triple-glazed, with thermal shutters. The entries will be rebuilt as air locks, preventing heat loss as tenants enter and leave the building. All of these changes will improve the building's thermal efficiency from around 30 Btu/DD/sq. ft. to 2.6 Btu/DD/sq. ft. If the government agencies administering appropriate funding programs should approve proposals, solar heat, supplied to the apartments by solar attics and a solar greenhouse, will provide up to 60% of the total heating requirement. Fans will draw the solar-heated air into the apartments.

Each apartment will store heat in phase-change salts placed above a false ceiling. These salts, when subjected to heat, change from a solid to a liquid condition, storing heat in this manner. As the salts give up the heat again, they change back to the solid condition. Phase-change salts provide an efficient, but relatively expensive, way to store heat. They are a desirable storage medium for this older building because of their light weight and lack of bulk, which enable them to be stored in a smaller space than rock or water would require.

While the building is being made more thermally efficient and adapted for solar heat, its living spaces will be restored to provide comfort for today's residents. The drafty, dark rooms of yesterday will be transformed into harmonious, light, bright, and cheery living areas, and the greenhouse will provide space for gardening.

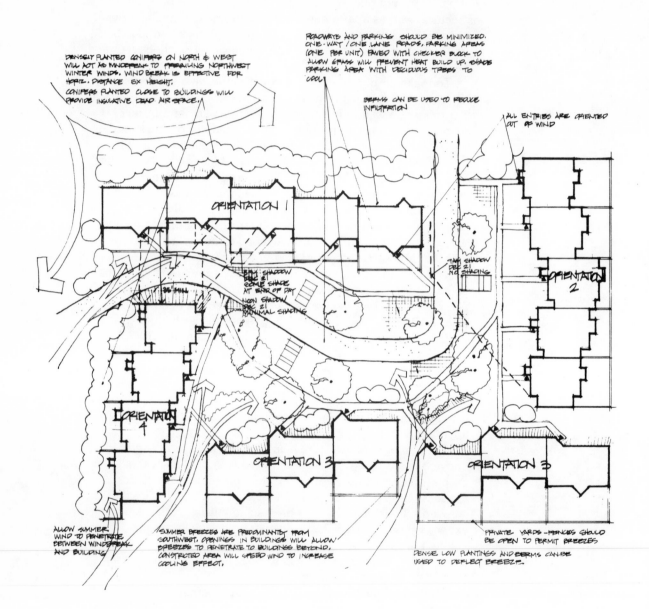

SITE PLAN

0 5 10   20   40   80

N

DENSELY PLANTED CONIFERS ON NORTH & WEST WILL ACT AS WINDBREAK TO PREVAILING NORTHWEST WINTER WINDS. WINDBREAK IS EFFECTIVE FOR HORIZ. DISTANCE 5X HEIGHT. CONIFERS PLANTED CLOSE TO BUILDINGS WILL PROVIDE INSULATIVE DEAD AIR SPACE.

ROADWAYS AND PARKING SHOULD BE MINIMIZED. ONE-WAY / ONE LANE ROADS. PARKING AREAS (ONE PER UNIT) PAVED WITH CHECKER BLOCK TO ALLOW GRASS WILL PREVENT HEAT BUILD UP. SHADE PARKING AREA WITH DECIDUOUS TREES TO COOL.

BERMS CAN BE USED TO REDUCE INFILTRATION

ALL ENTRIES ARE ORIENTED OUT OF WIND

ORIENTATION 1

ORIENTATION 2

9AM SHADOW DEC 21 NO SHADING

6PM SHADOW DEC 21 SOME SHADE AT END OF DAY

36" WALL

NOON SHADOW DEC 21 MINIMAL SHADING

ORIENTATION 4

ORIENTATION 3

ORIENTATION 3

ALLOW SUMMER WIND TO PENETRATE BETWEEN WINDBREAK AND BUILDING

SUMMER BREEZES ARE PREDOMINANTLY FROM SOUTHWEST. OPENINGS IN BUILDINGS WILL ALLOW BREEZES TO PENETRATE TO BUILDINGS BEYOND. CONSTRICTED AREA WILL SPEED WIND TO INCREASE COOLING EFFECT.

DENSE LOW PLANTINGS AND BERMS CAN BE USED TO DEFLECT BREEZE.

PRIVATE YARDS - FENCES SHOULD BE OPEN TO PERMIT BREEZES

Partial site plan for the
Milwaukee town house
project (1978).

There are 22 units to be reworked in this building, which is but one example of brand-new/old housing in an urban setting.

Where do such rehabilitation projects start? They start with an understanding on the part of owners, contractors, and city building officials that retrofit for solar heating is only construction—nothing new, nothing complicated. It is just a very sensible way of making a building cozy and so thermally efficient that little heat is required, compared to its former existence as an energy-wasteful building.

Other buildings similar to the one in Chicago can be found in all large cities in the Midwest and the East. If they are structurally sound, they can be rehabilitated, and if the sun strikes the building, it can be used to heat the apartments. How much of the total heating requirement can be supplied with solar heat will depend on the amount of surface available for the solar-heating system, compared with the amount of space that must be heated.

At the Chicago building, the sun strikes the building at roof level and on the upper part of the south wall at the level of the top floor, and there is sufficient roof space for solar attics and a solar greenhouse. At another location, you might find a building with solar access to a south-facing masonry wall—just waiting to be made into a mass heat wall. Or perhaps a wall can be opened up with enough windows for direct-gain heating.

Some buildings, especially the taller ones, may not present enough surface for solar retrofit to heat all of the apartments. In these cases, perhaps the upper units or the south-facing units can be solar heated, with a conventional system for the remaining units.

And some buildings will not have solar access at all, or they will be in such poor condition that they cannot be repaired. The redevelopment of city land is a continuous process, as buildings burn down, deteriorate, or are abandoned. It creates the opportunity to plan an entire site for solar access. Within the site, it is possible to provide solar heat for every residence, either single-family houses or apartment buildings (even high rises), while preserving solar access for adjoining property.

Planning New Developments. In 1978 a study funded by the AIA Research Corporation gave The Hawkweed Group an opportunity to plan an entire urban site for solar access, on a hypothetical basis. The purpose of the study was to find strategies for energy conservation and passive solar use. The site for our study was Milwaukee, while 23 other architectural teams worked on similar hypothetical projects in other parts of the country.

This study demonstrated that it is possible to build 20

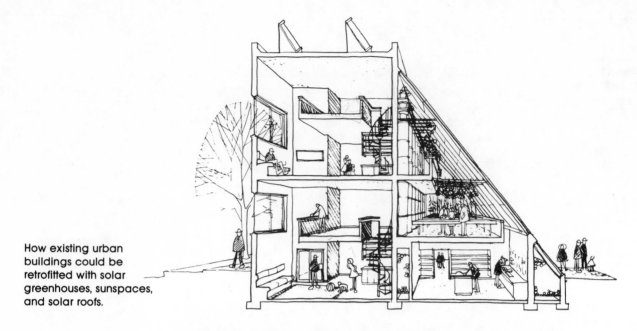

How existing urban
buildings could be
retrofitted with solar
greenhouses, sunspaces,
and solar roofs.

solar-heated town houses on a one-acre city parcel, using all of the
techniques to minimize heat loss that we have described in this book.
Even on such a small piece of land, one-third the size of a city block,
there is enough space to design appropriate responses to the climate
and to orient all of the buildings for south sun.

We put a lot of care into the Milwaukee town house project,
trying every idea that might help to reduce energy use in the houses.
And we looked for natural solutions for all energy-use areas.

Some of the town houses were designed with an east-west
orientation, while others had a north-south orientation. This was a
requirement of the project, and we were able to show the energy
benefits of facing solar houses toward the south. Those with an
east-west orientation used mass heat walls for solar heat; the others
had solar roofs.

We planned the site to protect the buildings from winter winds
and to let in cooling summer breezes. The entries were oriented to
the south and east, away from cold winds. Hard surfaces such as
parking lots and roads were avoided where possible, because they
store summer heat and radiate it to the houses and play areas. We
designed the landscaping for further protection from winter winds,
and for contribution to summer cooling. But that's not all. We also
looked for plant materials that were well adapted to the region, so
they could be easily maintained. We proposed the use of fruit trees
and the provision of space for vegetable gardens for a reduction in
energy use in the food distribution chain.

Recognizing the need to be concerned with all aspects of energy use, we planned for natural daylighting and for natural ventilation.

Having gone this far, we decided to see what else could be done. Hot-water heaters are high energy users, so a solar water heater would conserve energy. Furthermore, the conventionally powered backup water heater could be near the point of use—at kitchen and bath—and could be smaller than many now in use; no need to keep 80 gallons of water hot for a 10-gallon job of dishwashing. Flow restrictors on faucets and low-flush toilets would reduce water use, conserving energy needed at the community level for pumping and purifying the water supply and for sewage treatment.

Finally, we compared the energy requirements of units with an east-west orientation and units with a north-south orientation and found that east-west units would need only 8% of their total energy budget for heating, while north-south units would use 28% of their budget for heating. These north-south oriented buildings, with a reduced solar window, would use an additional 8,000 Btu's a year for each square foot of heated space. In either case, the total energy requirement is low, compared to conventionally planned buildings.

Building a Solar Town

Our Milwaukee study, showing what is possible in solar developments, was based on the underlying assumption that nothing would block the sun's access to the site. This, unfortunately, is a dangerous assumption today, since very few communities guarantee solar rights. It is of little use to redevelop the nation's cities as solar cities if a neighbor can put up a building or plant a tree that steals your sun. We alluded to this problem in our discussion of site selection in Chapter 5.

Your right to a clear solar window will only be assured when protective legislation is passed in all communities. Davis, California, and Soldiers Grove, Wisconsin, are leading the way here, with very clear ordinances. The Soldiers Grove requirement, part of a Planned Unit Development District Application, states: "Buildings shall not be of a height which would cast a shadow during daylight (hours) of 9:00 am and 3:00 pm of the winter solstice on any portion of another building or parcel if no building exists. Compliance with this standard must be graphically shown in Conditional Use applications for all parcels."

In this development you can safely build a solar building, assured of continued sunlight. The village itself controls landscaping and will take care to keep plants below the solar apertures. And Soldiers Grove has gone even further, regulating thermal efficiency by ordinance. Buildings are not permitted to have a heat loss greater than 4 Btu/DD/sq.ft.

This village of 500 residents is leading the way to the solar future. Faced with the need to relocate their business center out of a floodplain, the residents decided to build a new solar town center on a 19-acre site at the edge of the village. The new town will include municipal buildings, a post office, and 20 or more business structures. A grocery store, offices, a medical center, a craft shop, a service station, and a restaurant are among the sales and service establishments that are being constructed, and each of these buildings will be independently owned.

As architects and planners for the village, we are planning the project site and designing the solar-heated municipal buildings, which include a fire station to house rescue and fire fighting equipment for the volunteer fire department; a maintenance building for municipal equipment and work space; and a community building, with space for village offices and meetings, community activities, senior-citizen gatherings, and the library. We are also designing buildings for businesses relocating to the new town center.

The new town center of Soldiers Grove, Wisconsin. All the buildings are sited facing south and are far enough away from the mountains to receive full sunlight all year round.

Planning for the Soldiers Grove project began with the village. Businessmen, village officials and staff, and many interested

A cardboard model of the new town center of Soldiers Grove was put on display at the Community Development Office, for review by local citizens.

residents were involved in the project from its infancy. With help from state agencies, they developed a conceptual plan to guide us in our work. However, more precise topographical information forced some adjustments of their concept plan.

Recognizing that solar access might be blocked by the mountains to the south, we made studies to define the solar window. Using a transit-like instrument mounted on a tripod, called a solar site selector, we were able to ascertain that the entire 19-acre site would be in shadow beginning at 2 pm on the winter solstice. For the crucial time when heat is needed most, there would not be enough sun. So we modified the plan, moving the building sites northward into the sunshine.

A large-scale drawing and a cardboard model at the same scale helped us to determine the actual relationships of the buildings to each other and also enabled everyone involved to understand how the new town center would look. The model was put on display in the Community Development Office to provide a graphic answer to residents' questions.

At the same time that we were planning the solar town center for Soldiers Grove, the village was investigating the possibility of

One of the first buildings at Soldiers Grove nearing completion. Heated by a solar attic, it will house a grocery store.

using a wood-fueled central heating plant to provide backup heat for all of the buildings, circulating either steam or hot water, since wood is locally available as a by-product of lumbering and manufacturing. However, it turned out that the plant would not be economically feasible for buildings as thermally efficient as those that we were designing. The village had asked for high thermal efficiency and passive solar design. We had obliged. They had embraced our proposed standards readily and encouraged us all the way. As a result, so little backup heat was needed that building a central heating facility was unnecessary. Instead, backup heat for each building will be provided by a propane-fired hot-air furnace; eventually the propane will be replaced with locally manufactured methane.

Not only did we plan the buildings for high thermal efficiency, we also designed the landscape for climate control. The principles developed encouraged earth berms, windbreaks, and use of plants for evaporative cooling. In addition, we inventoried trees and plants on the property and designed a landscape plan around them, rather than moving plants to the site at a high energy cost and maintaining exotic plants. We found white and red pine, wild plum, gray dogwood, asters, Solomon's seal, wild strawberry, wild grapes, and many other fine trees and plants to use in the landscape. Large evergreens on the site will be moved to establish an instant windbreak.

The buildings we have designed for the project exhibit some common characteristics. They are frame buildings, very thermally efficient, and they receive most of their heat from the sun—with site-built passive systems.

The supermarket is an all-wood building, with a concrete

foundation. The wood has been treated to withstand exposure, so we can put earth up against the walls on all sides to reduce heat loss by infiltration. We used 2-by-6-inch framing at 24 inches on center, and there are 6 inches of batt insulation in the walls—sufficient for this kind of building with its lower temperature requirements. The columns are wood and the beams are laminated wood. The roof is pitched at a 4/12 slope (a 4-foot drop in elevation for every 12 feet of roof) to the public space north of the building to help prevent snow buildup in this northern climate.

The south portion of the attic, with 16-foot glazed panels along the entire length of the building, forms a sunspace, admitting the sun into a black-painted space. Heated air flows through ductwork that also serves the backup heating system. The same blower fan is used to move solar heat and backup heat. Heat from the refrigeration compressors is reclaimed for space heating, also. Close to 100% of the building's heat will be provided by the combination of solar attic and heat reclamation, according to our calculations.

Interior of the solar attic of the grocery building, Soldiers Grove.

Medical center

| | X - RAY | DK RM | LOUNGE | CONSULTATION | SHOWER | NURSE/LAB | MECHANICAL | STORAGE | CONSULTATION | M | W | WAITING |

AIR LOCK

CORRIDOR

HALL

| EMERGENCY EXAM | EXAM 4 | EXAM 3 | EXAM 2 | EXAM 1 | TEACH | OFFICE | RECEPTION | ENTRY |

First Floor Plan

N
0 2 4 6 8 10

We investigated various heat-storage methods. One day, picking up a can of soup for lunch, we realized that the storage mass was already there—a normal part of every supermarket. Every can of soup, every bottle of juice—in short, all of the food stored on the shelves—is a thermal mass (sun for sale), holding low-temperature heat. Not hot enough to affect the condition of the canned goods, it is still useful heat for maintaining an even temperature in the building.

Drywalled, white-painted interior surfaces will reflect light and lessen the need for artificial illumination. Further energy savings will be realized with fluorescent lighting, which uses the least energy of any lighting system.

Solar supermarkets, tastefully furnished with shelves of canned and boxed goods, may someday be the norm. Today, a grocery store owner in a village of 500 is ahead of the times.

Another of our projects at Soldiers Grove, also under construction as we write, is a medical center. About 2,800 square feet in size, it has concrete footings and frame walls, in keeping with project standards. Twelve-inch-thick walls, actually a pair of spaced 2-by-4 stud walls, hold 12 inches of insulation, with 24 inches in the ceiling. The windows are triple-glazed except on the south, where they are double-glazed. Heat is provided by a solar attic similar to that at the supermarket, and it is stored below the concrete floor in a system of sand and under-slab ducts. Heated air flows from the storage bed or the attic as needed, propelled by a blower shared with the backup heating system.

As in the supermarket, the interior surfaces will be drywall painted in light colors to reduce the need for artificial lighting. Windows in every room provide daylight and open for ventilation. The entrances to the building have air locks, to reduce infiltration heat loss.

A third project, a rental office with an apartment above, is similar in construction to the medical center. It has windows only on the south, double-glazed for direct gain and covered with thermal shutters at night.

Several solar-heating strategies are combined in this building. There is direct gain on both the first and second floors and a mass heat wall at the carport, with a small fan to deliver heat to the first floor. A solar attic at the roof augments second-floor heat.

Heat storage is also varied. The mass heat wall stores its own heat. Direct-gain heat for the first floor is stored in the rental office floor slab, while heat for the second-floor apartment is stored in water bottles above the stairway.

All three buildings have thermal shutters to cover the windows at night. All are tightly built, thoroughly weather–stripped, and carefully caulked.

Let's talk about construction prices. While we were making studies, developing standards, and preparing working drawings for these buildings, there was much discussion around town about costs, with all kinds of guesses and worries thrown about. Our insistence that these buildings would cost the same as nonsolar buildings being constructed in the area may have been reassuring, but nothing beats positive proof.

Only bids would resolve the debate. And when the drawings were complete, we were ready to solicit bids. Our drawings went to 75 contractors, representing all of the trades needed. All of the

The Schoville Building, Soldiers Grove, under construction. The multi-use building combines a mass heat wall at the carport, direct-gain windows on the first and second floors, and a solar attic.

The Schoville Building

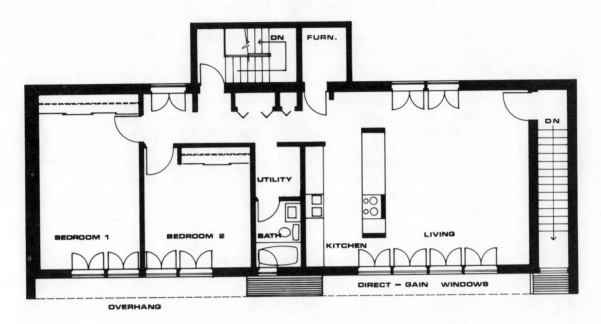

DN
FURN.

DN

UTILITY

BEDROOM 1
BEDROOM 2
BATH
KITCHEN
LIVING

DIRECT — GAIN WINDOWS

OVERHANG

Second Floor Plan

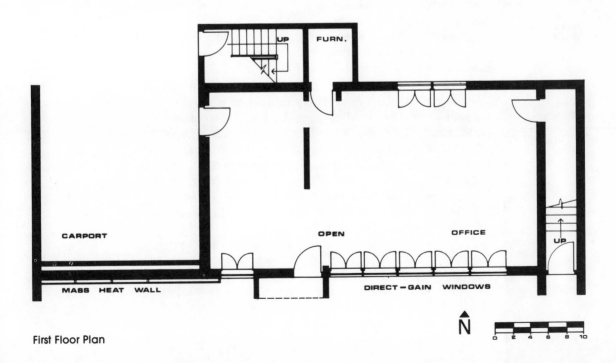

UP
FURN.

CARPORT
OPEN
OFFICE
UP

MASS HEAT WALL
DIRECT — GAIN WINDOWS

N

0 2 4 6 8 10

First Floor Plan

contractors were from the area. Few knew anything about solar heating. After about a month of intensive activity, with phone calls and visits to our office, bids were turned in for all of the necessary trades for the three buildings. As we had predicted, the prices were like those for nonsolar buildings in the area. In each case, the building prices were below the budget.

So a group of contractors, selected for their capability to do the job, began work on their first solar buildings, not in the least inhibited or hindered by their lack of experience in the solar field.

The construction process has proved no more complex than the bidding. As a client once told us, "I thought I would have problems with the solar parts, but I didn't. The problems I had were the same old ones—men not coming to work on Monday morning, materials not arriving on time—those kinds of things." The same can be said of Soldiers Grove.

The exciting project at Soldiers Grove is leading the way toward the solar future. Nowhere else in the country is a community committing itself so personally to solar heating. The example Soldiers Grove residents are offering, the experience they are having, the learning that is going on there will help communities across the country move into the solar world.

The city of Davis, California, has chosen a different way, developing energy conservation and solar strategies, educating residents to the advantages of using conservation and solar energy, and, finally, adding incentives for these strategies to the city subdivision requirements.

Not all communities are as prepared as Soldiers Grove and Davis to use the sun, or even to encourage its use. A typical city plan commission is not concerned about thermal efficiency, solar orientation, sun rights, or energy conservation. For example, a 27-unit residential development of our design is moving slowly through the formalities of approval by a suburban plan commission. At the meetings, the developers discuss solar access and thermal efficiency, while the plan commissioners discuss parking spaces. Gradually, however, mutual understanding is developing, and the project will probably soon be approved.

Designed to be as low in energy demand as possible, it includes gardens, orchards, and pedestrian paths to nearby schools and shops. The project was self-developed by a group of families who want to control their own housing and share their commitment with others, and it is strongly oriented toward self-sufficiency for its owners.

Town houses will be passively solar-heated and will share

greenhouses and workshops. The buildings will be tightly clustered, in order to free land in the nine-acre site for vegetable gardens and play areas for children and adults.

Throughout the project's life, all of us have moved together, each understanding the others better as we progressed from step to step. And the marvelous result of this project will be 27 town houses that are truly planned by their owners, who will certainly understand very well what they have chosen.

What Lies Ahead?   What do all of these projects tell us of the future, of the solar age that our nation is entering? We think that this country is on the verge of an era of true conservatism, an age when it will choose to be regional in its resource management and use. We won't be looking for solutions woven of moonshine and fancy, but for realistic solutions—simple, careful, painstaking—in our backyards and on our rooftops.

The solar solution is already becoming a reality, spreading by word of mouth, by community college seminars and workshops, by hundreds of hammers and saws applied to hundreds of home-built solar collectors and greenhouses. And each of these solar buildings is an important step into the solar future.

What we must do now is to tell everyone about solar heating—all of the people who haven't heard or aren't sure. One way is through alternative energy organizations to disseminate information on a community-wide basis. Another possibility is county energy agents, similar to the county agents who brought better farming practices to the country during and after the Great Depression. The county energy agents would be men and women knowledgeable about solar heat, windmills and wind generators, organic farming methods, wood-burning stoves and furnaces, woodlot maintenance, and the production of alternative fuels such as alcohol and methane.

These agents would visit you at home, at the PTA meeting, or wherever you gather. At the neighborhood level, they would organize the change to alternative energy and help you to make it happen. Serving as advisors to local advisory groups, commissions, and elected officers, the agents would also help to write new codes and zoning ordinances. These new regulations would protect your solar rights, enabling you to proceed with confidence to build a solar house, knowing that the sun would not be blocked at a later time.

The large cities are a special case—and an important one. In the interest of energy conservation, the existing housing should be retained whenever possible. To tear down these buildings, make new materials, and build anew would be an energy-intensive process—one we cannot afford.

The creative re-use of these buildings, solar-heated and thermally efficient, will renew the cities and make them highly desirable places in which to live. And, as we rehabilitate the cities, we can work for a greater degree of self-sufficiency. Greenhouses as heat sources and food providers will offer city folk a chance to share the benefits that rural and small-town people are already experiencing. You who live in the cities will be able to grow enough food to give you salads from your own greenhouse, breaking the cycle of dependence upon long production chains. With urban gardens in yards and vacant lots, you can grow your own summer vegetables, setting some food aside for winter. Already, in Chicago and other cities, tomato plants and radishes and zucchini are replacing the petunias and pansies in borders and along fences.

The most important single thing for us to recognize is that we don't really need help from government agencies, county energy agents, or anyone at all to get started. The county agents might make the process move a little faster. The government agencies might help or they might slow us down, who knows?

Solar heating, gardens, resource conservation—all of it is right here, right now, waiting for us. All we need to do is use it.

One of the raised-bed gardens at The Hawkweed Group's farm near Osseo, Wisconsin.

Glossary

Active solar system: A high-technology solar-heating system that relies on mechanical aids for distributing heat.

Backup system: A supplementary heating system that provides heat on cloudy or extremely cold days, when your solar system cannot satisfy your heating requirements.

Berm: A mound of earth placed against an outer wall of a building, to help prevent heat loss in winter and heat gain in summer. *See also* Earth integration.

Btu (British thermal unit): The amount of heat required to raise the temperature of 1 pound of water 1° Fahrenheit.

Conduction: A natural process in which heat is distributed by passing from molecule to molecule within an object, or between two touching objects. The rapid vibrations of heated molecules cause adjoining molecules to vibrate.

Conduction heat loss: Loss of heat that occurs when heat passes through the exterior elements of a building—walls, windows, roof, doors, floors—to the cold outside. *See also* Conduction.

Convection: A natural process in which heat is transmitted from a warmer to a cooler surface through the movement of air. Heated air expands and rises, distributing its heat throughout a room.

Degree Day (heating): A measure of heating need determined by subtracting an average daily temperature below 65° Fahrenheit from the base number 65. A day with an average temperature of 50° has 15 Degree Days.

ΔT: A figure used in calculating heat loss. It is determined by subtracting the design temperature from the indoor thermostat setting. *See also* Design temperature.

Design temperature (for heating): The minimum above which the outdoor temperature will remain 97.5% of the time. You can find the design temperature for your area by consulting the ASHRAE *Handbook*.

Direct gain: A passive solar-heating method in which sunlight enters a building through south-facing windows and strikes a heavy building component, such as a concrete floor slab or a masonry wall, which stores the solar heat. The heat is slowly released to warm the room.

Earth integration: Building a structure partially or completely underground, except on the south, to help prevent heat loss in winter and heat gain in summer. *See also* Berm.

Glazing: A transparent or translucent material, usually glass or fiberglass, that freely transmits sunlight. Glazing is used in passive solar-heating systems to admit sunlight into a confined space, where it is converted to heat.

Glazing load ratio (GLR): A building's heating requirement, expressed in Btu's per Degree Day, divided by the glazing area, expressed in square feet. The GLR tells you what percentage of your annual heating requirement can be met by solar contribution.

"Greenhouse effect": The ability of glass and other transparent materials to transmit sunlight into an enclosed space and to trap the resulting heat.

Infiltration heat loss: Loss of heat that occurs when cold outside air enters a building through cracks and must be heated.

Internal heat gain: Heat generated inside a building by its occupants, by cooking, and by lighting and electrical appliances.

Macroclimate: The overall weather conditions in a given area.

Mass heat wall: A south-facing wall of a building, made of concrete, concrete block, brick, or stone and covered on the outside with three separated layers of glazing, or two layers and nighttime insulation, that collects and stores solar heat. The wall radiates heat into the building and creates a flow of warm air that enters the building through vents. It is also known as the Morse Wall and the Trombe Wall.

Mesoclimate: The weather conditions at a specific building site. Mesoclimate is affected by such factors as proximity of woodlands or a body of water, and location on a slope or in a valley.

Microclimate: The climatic conditions within a building, such as temperature and relative humidity.

Passive solar system: A low-technology solar-heating system that is architecturally integrated (the building itself is the heating system) and relies only minimally, if at all, on mechanical aids for distributing heat.

Peak heat load: A building's maximum rate of hourly heat loss on a cold night, when the outdoor temperature is the same as the design temperature. *See also* Design temperature.

Phase-change salts: Salts used for heat storage. They absorb heat, melting in the process, and release heat as they solidify again.

R Value: A measure of a given material's resistance to heat loss. R Values for various materials can be found in the ASHRAE *Handbook*.

Radiation: A natural process in which heat is transmitted from a warmer object to a cooler one by waves of radiant energy given off by moving molecules on its surface. When the waves of energy strike the cooler object, they heat it by speeding up its molecules.

Reflectance: The percentage of the total amount of light striking a surface that is reflected by it.

Retrofit: Re-fitting an existing building for solar heat.

Solar attic: A sunspace built above the wall line on the south side of a building, as part of the roof configuration. It requires a fan for heat distribution, and may store heat in a remote rock bed. *See also* Sunspace.

Solar radiation: Short-wave radiant energy given off by the sun.

Solar roof: A section of a south-facing roof with two separated layers of glazing over black-painted corrugated metal. The metal, heated by the sun, heats the air around it, which is removed by a fan and propelled to a remote rock bed for storage.

Solar window: A projection of the paths of the sun across the sky at different times of the year, in relation to a given building. The solar window is useful for determining when the building will be in shade.

Stack effect: A flow of cool air into a building through inlet windows facing the prevailing winds, forcing heated air out of the building through high vents in the opposite wall. It is used as a natural ventilation technique.

Summer solstice: The day of the year, around June 22, when locations in the Northern Hemisphere receive the most daylight because of the angle of the sun.

Sunspace: A glazed space along the south wall or roof of a building that traps solar heat, which is transmitted to the interior of the building via convection currents or fans. A sunspace might be a greenhouse or a solar attic.

Thermal efficiency: A quality of a building in which there is minimal heat loss in winter and heat gain in summer because of thorough insulation and tight construction, making possible a lower expenditure of energy for heating and cooling than in an energy-wasteful building.

Thermal mass: A heavy building component, such as a concrete floor slab or a masonry wall, used for heat storage.

Thermal radiation: Heat. Solar radiation is converted to thermal radiation when it strikes an object.

Thermal shutters: Insulated, tightly fitting shutters that are closed over windows at night to reduce heat loss.

Trombe Wall: *See* Mass heat wall.

U Value: A measure of heat loss through a particular building component when outside air is 1° Fahrenheit colder than inside air, expressed in terms of Btu's per square foot per hour. The ASHRAE *Handbook* lists U Values for some building components.

Vapor barrier: A plastic film that is stapled to the studs of a building under construction to prevent the passage of moisture from the interior of the building to the framing members.

Winter solstice: The day of the year, around December 22, when locations in the Northern Hemisphere receive the least daylight because of the angle of the sun.

Bibliography

These are some of the books on our shelves that we turn to when we work with solar houses.

For background information about climate, we have found the following books useful: Rudolf Geiger's *The Climate Near the Ground,* translated by Milroy N. Stewart and others (Cambridge, Massachusetts: Harvard University Press, 1950), and Glenn T. Trewartha's *An Introduction to Climate,* Fourth Edition (New York: McGraw-Hill, 1968).

The National Oceanic and Atmospheric Administration (NOAA) of the U.S. Department of Commerce is an invaluable source of climatic data. Among its publications are *Climatic Atlas of the United States* (published in 1968 and revised in 1977), *Local Climatological Data* for cities with weather stations throughout the United States (published annually), and *Comparative Climatic Data for the United States Through 1978.* You will find these publications referred to throughout this book, and you may want to have them on your own shelf.

For a discussion of the relationship between climate and building design, we recommend Victor G. Olgyay's *Design with Climate: Bioclimatic Approach to Architectural Regionalism* (Princeton, New Jersey: Princeton University Press, 1963).

The American Society of Heating, Refrigerating, and Air-Conditioning Engineers (ASHRAE) publishes some handbooks that will be helpful when you are calculating your solar house's thermal efficiency and the amount of glazing required to heat it. These are the ASHRAE *Handbook of Fundamentals* (New York, 1972), and the ASHRAE *Handbook and Product Directory, 1974: Applications* (New York, 1974).

Finally, you might want to consult *Passive Solar Design Analysis* (1980), which is the second volume of *Passive Solar Design Handbook.* Prepared for the Department of Energy, it is available through the National Technical Information Service (NTIS), which is part of the Department of Commerce.

Index